# STRUCTURE OF MOLECULES
## AND
# INTERNAL ROTATION

# PHYSICAL CHEMISTRY
## *A Series of Monographs*

### Edited by

## ERIC HUTCHINSON

*Stanford University, California*

W. Jost: Diffusion in Solids, Liquids, Gases, 1952

S. Mizushima: Structure of Molecules and Internal Rotation, 1954

H. H. G. Jellinek: The Degradation of Vinyl Polymers, in preparation

## ACADEMIC PRESS INC., PUBLISHERS

# STRUCTURE OF MOLECULES

## and

# INTERNAL ROTATION

*By*

SAN-ICHIRO MIZUSHIMA

*Professor of Physical Chemistry of the University of Tokyo*

1954

ACADEMIC PRESS INC., PUBLISHERS

NEW YORK

Library of Congress Catalog Card Number: 55—11054

*Cat for Chemistry*

Dedicated to

DR. MASAO KATAYAMA

*Emeritus Professor*
*of the University of Tokyo*

# Preface

For several years I have been planning to write a book on internal rotation, a subject in which I, with many coworkers, have been interested during the past twenty years.

The opportunity and incentive to publish this book in English have been prompted by the kind invitations I have accepted to serve as a visiting lecturer and to speak at some symposia. From 1951 to 1953 requests were received from several universities and institutions in the United States and in Europe, including the University of Notre Dame, Cornell University, and Consejo Superior de Investigaziones Cientificas. My decision to write the book at this time has been encouraged by Professor E. Hutchinson, who wished to include this book in his international series. I should like to express my warmest gratitude to Professor P. Debye, Sir Lawrence Bragg, and Professors C. C. Price, O. R. Foz Gazulla, J. W. Williams, F. A. Long, G. Glockler, and P. C. Cross for their cordial invitations. Thanks are also due to Professor P. A. McCusker of the University of Notre Dame for some valuable suggestions.

The first part of this book deals with a description of the development of the investigations on internal rotation. In the second part are presented a more detailed explanation of some of the theoretical aspects of the problem, especially that of normal vibrations, and a description of experimental methods. This is built up as systematically as possible so that this may be understood by the chemists whose special field is in other directions but who wish to increase their knowledge of the general field of structural chemistry. Therefore, it was necessary to include some sections well known to the specialists.

I am deeply indebted to the University of Tokyo for the extension of leave which permitted me to visit the universities and institutions on both sides of the Atlantic and to attend some international congresses and symposia for valuable discussions on the subject matter of this book. I also wish to thank many coworkers of mine, including Professor Y. Morino and Dr. T. Shimanouchi, who have been working with me on this subject for so long, and Drs. M. Yasumi, I. Watanabe, I. Ichishima, S. Nagakura, K. Kuratani, and M. Tsuboi, and Messrs. T. Miyazawa, T. Sugita, and I. Nakagawa for their helpful discussions in preparing this book.

*Tokyo*, 1954.

**S. Mizushima**

# CONTENTS

## Part I

# Part II

# PART I

# CHAPTER I

## Introduction and Earlier Investigations

### 1. Introduction

The structure of molecules was originally studied by chemical methods of investigation involving the determinations of the composition of substances, molecular weights, the existence of isomers, the nature of the chemical reactions in which substances take part, and so on. From the consideration of facts of this nature chemists succeeded in determining not only the structural formula, but also the spatial configuration of molecules: e. g., van't Hoff and le Bel were led to bring classical organic stereochemistry into its final form by their postulate of the tetrahedral valences of the carbon atom.

Perhaps the most striking advance of the present century in the field of physical chemistry is the more general and quantitative elucidation of such stereochemical problems by the application of various physical methods, including the study of the structure of crystals by the diffraction of X-rays and of gas molecules by electron diffraction, the measurement of electric dipole moments, the interpretation of infrared and Raman spectra, and the determination of entropy and specific heat values.

Among the many interesting results obtained by these methods, that relating to interatomic distances is worthy of note. The values found for the equilibrium distance between two atoms A and B connected by a covalent bond of fixed type in different molecules and crystals are in most cases very nearly the same, so that it becomes possible to assign a constant value to the A – B bond distance for use in any molecule involving this bond. For example, the carbon-carbon distance in diamond (representing a single covalent bond) is 1.542A, and the values found in many molecules containing this single bond are equal to the diamond value to within their probable errors. This constancy is of interest in view of the varied nature of the molecules. Similar constancy is shown by other covalent bond distances.

Moreover, it is found, that covalent bond distances are in general related to one another in an additive manner: the bond distance X – Y is equal to the arithmetic mean of the distances X – X and Y – Y. For example,

3

the C – C distance is 1.54A, as stated above, and the Cl – Cl distance in the element is 1.98A. The arithmetic mean of these, 1.76A, is identical within the probable error of the experimental value with the C – Cl distance found in many organic molecules containing this bond. In consequence it becomes possible to assign to the elements *covalent radii* (shown in Table 1.1) such that the sum of two covalent radii is equal to the equilibrium internuclear distance for the two corresponding atoms connected by a covalent bond.

TABLE 1.1

Covalent radii*

| Atom | Single bond | Double bond | Triple bond |
|------|-------------|-------------|-------------|
| H | 0.30 | | |
| B | 0.88 | 0.76 | 0.68 |
| C | 0.77 | 0.67 | 0.60 |
| N | 0.70 | 0.61 | 0.55 |
| O | 0.66 | 0.57 | |
| F | 0.64 | | |
| Si | 1.17 | 1.07 | 1.00 |
| P | 1.10 | 1.00 | 0.93 |
| S | 1.04 | 0.95 | |
| Cl | 0.99 | | |
| As | 1.21 | 1.11 | |
| Se | 1.17 | 1.08 | |
| Br | 1.14 | | |
| Sn | 1.40 | 1.30 | |
| Te | 1.37 | 1.28 | |
| I | 1.33 | | |

Thus one might expect to be able to construct any molecular model, even that of a polymer, by using the bond radii and known bond angles. There is, however, a further problem in structural chemistry remaining to be solved: i. e. the question of internal rotation about a single bond as axis, which is the subject of this book. Without the knowledge of the internal rotation we cannot even determine the structure of such a simple molecule as ethane, $H_3C–CH_3$.

---

* These values are taken from L. Pauling, The Nature of Chemical Bond, Cornell University Press (1939). See also V. Schomaker and D. P. Stevenson, J. Am. Chem. Soc., **63**, 37 (1941). These authors have suggested that corrections should be made for the additivity rule, when the covalent bond acquires ionic character: in other words, when the two atoms forming the bond differ very greatly in their electronegativity.

Such internal rotation about a single bond as axis was for some time supposed to be entirely free and even the term "free rotation" has often been used in chemistry instead of the term "internal rotation" now generally accepted. We must, however, be clear as to what the so called "principle of free rotation" originally meant. It was founded upon the non-existence of isomers whose structures could pass into one another by means of the internal rotation about a single bond: it did not assume that the rotation went on continuously in the molecule, or that all the positions to which it could lead were equally favorable. Actually considerable evidence based on the various physical methods referred to above has been accumulated to indicate that the rotation is not free and that there is in general more than one potential minimum in one complete rotation about a single bond as axis. The rotational isomers corresponding to these minima differ only by small amounts in their properties and any attempt to separate them by the usual chemical methods will not succeed except in some special cases.

The most interesting example of such a special case is afforded by the diphenyl derivatives with three or four substituents in the 2,2′, 6,6′ positions, such as

OCH$_3$   COOH         CH$_3$  F              NH$_2$   CH$_3$

HOOC         OCH$_3$        CH$_3$  NH$_2$          F     CH$_3$
     (I)                                    (II)

Compounds of this type have been extensively studied by Adams and co-workers.[1] The substituents in the 2,2′, 6,6′ positions restrict, by their spatial interference, the free rotation of the two benzene nuclei around the common axis, thus preventing the rings from becoming coplanar and thereby producing in the molecule an asymmetric configuration. It was found possible by properly modifying the size of the 2,2′, 6,6′ groups to prepare optically active diphenyls with widely varying degrees of stability towards racemization. For example, for the two diphenyl derivatives whose formulae appear above, (II) is completely racemized after 30 minutes boiling in acetic acid, while for (I) the half-life period for boiling in the same solvent is 78 minutes.[1] The racemization of optically active diphenyl derivatives is explained in this case on the basis of the theory of restricted rotation, in that thermal

---

1. R. Adams et al., Chem. Rev. 12, 261 (1933). See also many other papers published in J. Am. Chem. Soc.

agitation causes the groups in 2,2', 6,6' positions to slip by each other (i. e. to cross over the potential barrier) and thus result in internal rotation of the two benzene nuclei.

The simplest molecule with an axis of internal rotation is ethane, $H_3C-CH_3$ for which we can expect a potential curve of the type shown in Fig. 1.1. The potential curve can be obtained theoretically, if we know the interaction of the H atoms of one methyl group with those of the other which prevents free rotation about the C–C axis. Assuming a simple type of interaction energy, Eyring et al. calculated the height of the potential barrier to internal rotation in ethane as 0.3 kcal/mol.[2]

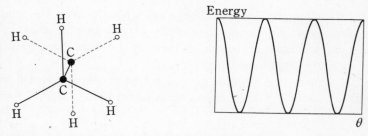

Fig. 1.1. The dependence of the internal potential of ethane on azimuthal angle.

According to classical mechanics, the specific heat of a gas can be calculated by allowing to each degree of freedom of the molecules of gas a mean energy $RT/2$. Thus the specific heat is $nR/2$, where $n$ is the number of degrees of freedom and $R$ is the gas constant, with a value approximately 2 cals. per mole. If ethane were a rigid body, it would have three degrees of translational and three degrees of rotational freedom, and so the value of $n$ would be 6. There is, however, an additional degree of freedom, the twisting of one methyl group relative to the other. This motion is clearly on a different footing from the other six, because it can be inhibited at temperatures still high enough to permit considerable translational and rotational activity. One would expect, therefore, that at low temperatures the specific heat would be $3R$, and that at higher temperatures it would approximate to $7R/2$. The temperature dependence of the contribution to the specific heat of the twisting motion can be calculated by quantum mechanics (see Section 10), if we know the form of the potential curve, which can be determined by the calculation of Eyring referred to above.[2]

---

2. H. Eyring, J. Am. Chem. Soc. **54**, 3191 (1932).

The result of this calculation[3] was in good agreement with the experimental material of Eucken and Weigert[4] and thus the most fundamental problem of the internal rotation seemed to be solved.

However, as we shall see in Section 10, a more accurate determination of specific heat has required the height of the potential barrier to be about 3 kcal./mole which is about ten times as large as that obtained by Eucken and Weigert. We have to consider, therefore, that the calculation of the interaction energy between the two methyl groups of ethane is far more complicated than would appear at first sight. Such being the case, it is desirable to choose other kinds of molecules, for example, 1,2-dihalogeno-ethane, $XH_2C-CH_2X$, for which the interaction between the two halogen atoms might play a much more important part than that between two hydrogen atoms or that between hydrogen and halogen atoms. This seemed quite reasonable in the earlier stages of the investigation of the internal rotation, since as stated above, the interaction between two hydrogen atoms of different methyl groups of ethane was considered to be small.

Therefore, many researchers considered at that time that the internal rotation of 1,2-dihalogenoethanes could be discussed by means of a simple potential function with one potential minimum at the position at which the two halogen atoms are at the farthest distance apart. From the experimental point of view the halogenated ethanes have several advantages over ethane itself: for example, we can apply the dielectric measurement which is the simplest experimental method for attacking the internal rotation problem of such a kind of molecule. Let us first discuss the results of such measurements on 1,2-dihalogenoethanes as earlier studies in this field.

## 2. Earlier Studies on Dihalogenoethanes

The molecular configuration of 1,2-dihalogenoethanes, $XH_2C-CH_2X$, with two like halogen atoms on different movable groups has a center of symmetry, when these two halogen atoms are at the greatest distance apart (see Fig. 1.2). We shall call this configuration the *trans* form and take the origin of the azimuthal angle $\theta$ of internal rotation at this position. It is evident that the dipole moment of this configuration is zero while that of any other configuration is finite. It is also evident that the value of the

---

3. E. Teller and K. Weigert, Nachr. Göttingen Ges., 218 (1933).
4. A. Eucken and K. Weigert, Z. phys. Chem. **B 23**, 265 (1933).

dipole moment increases with increasing angle of internal rotation, until it reaches the maximum at the *cis* position, in which all the atoms of one movable group eclipse the same kind of atoms of the other group as viewed along the C–C axis.

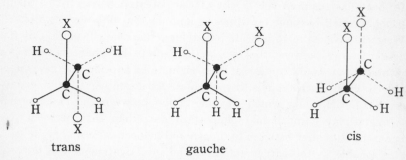

| trans | gauche | cis |

Fig. 1.2. The configurations of $XH_2C–CH_2X$ as viewed along the C–C axis.

According to the experimental results of the earlier investigations in this field[5—9], the apparent moment of 1,2-dihalogenoethanes increases with rising temperature (see Table 1.3). We can, therefore, consider that the *trans* form is the most stable form and as a result of increased thermal energy the molecule tends to take a configuration with larger azimuthal angle $\theta$ of internal rotation, as the temperature is raised.

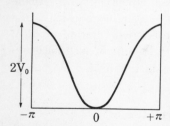

Fig. 1.3. A simple potential curve of internal rotation.

Let us first consider only the interaction between the two $C - X$ groups and assume a potential function for the internal rotation which has only one potential minimum at the *trans* position at which we take the origin of azimuthal angle $\theta$ (see Fig. 1.3).

$$V = V_0 (1 - \cos \theta) \qquad (1.1)$$

Here $2 V_0$ denotes the energy difference between the *cis* and the *trans* forms. Starting from such a simple potential function Mizushima and Higasi[9]

5. L. Meyer, Z. physik. Chem., **B 8**, 27 (1930).

6. C. P. Smyth, R. W. Dornte and E. B. Wilson, J. Am. Chem. Soc. **53**, 4242 (1931).

7. C. T. Zahn, Phys. Rev. **38**, 521 (1931).

8. E. F. Greene and J. W. Williams, Phys. Rev. **42**, 119 (1932).

9. S. Mizushima and K. Higasi, Proc. Imp. Acad. Tokyo, **8**, 482 (1932).

calculated the apparent moment of 1,2-dihalogenoethanes, $XH_2C-CH_2Y$, as follows:

Let $\mu_x$ and $\mu_y$ be the group moments of $CH_2X$ and $CH_2Y$, repectively, both of which make an angle $\alpha$ with $C - C$ axis (see Fig. 1.4). Then the dipole moment $\mu$ of a configuration* with azimuthal angle $\theta$ can be calculated as follows:

$$\mu^2 = (\mu_x - \mu_y)^2 + 2\mu_x\mu_y \sin^2\alpha\,(1 - \cos\theta)$$

The probability that a molecule has the moment value $\mu$ is given by the Boltzmann law as

$$e^{-\frac{V_0(1-\cos\theta)}{kT}}$$

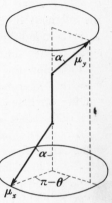

Hence we obtain as the average square moment of the molecule at temperature $T^0$K:

$$m^2 = \frac{\int_{-\pi}^{+\pi} \mu^2 e^{-\frac{V_0(1-\cos\theta)}{kT}}\,d\theta}{\int_{-\pi}^{+\pi} e^{-\frac{V_0(1-\cos\theta)}{kT}}\,d\theta}$$

Fig. 1.4.

$$= (\mu_x - \mu_y)^2 + 2\mu_x\mu_y \sin^2\alpha \left[ 1 - \frac{\int_{-\pi}^{+\pi} \cos\theta\, e^{-\frac{V_0(1-\cos\theta)}{kT}}\,d\theta}{\int_{-\pi}^{+\pi} e^{-\frac{V_0(1-\cos\theta)}{kT}}\,d\theta} \right]$$

If we put $-\dfrac{V_0}{kT} = i\,z$, then the integrals can be transformed into Bessel's form and we have:

$$m^2 = (\mu_x - \mu_y^{\cdot})^2 + 2\mu_x\mu_y \sin^2\alpha \left[ 1 + \frac{i\,J_1\left(i\,\dfrac{V_0}{kT}\right)}{J_0\left(i\,\dfrac{V_0}{kT}\right)} \right] \qquad (1.2)$$

where $J_1$ and $J_0$ denote the Bessel functions of the order 1 and 0, respectively, In the case $\mu_x = \mu_y = \mu_0$ this formula is reduced to

---

* This is the configuration in which the projections of two moments on a plane perpendicular to $C - C$ axis make angle $\pi - \theta$.

$$m^2 = 2\mu_0^2 \sin^2 \alpha \left[ 1 + \frac{i\, J_1\left( i\, \dfrac{V_0}{kT} \right)}{J_0\left( i\, \dfrac{V_0}{kT} \right)} \right]. \qquad (1.3)^*$$

To obtain a theoretical value of $m$, it is necessary to know the values of $\mu_x$, $\mu_y$, $\alpha$ and $V_0$. In the first approximation we can put $\mu_x$ or $\mu_y$ equal to the moment of the corresponding methyl halide, its direction coinciding with the bond direction linking carbon to halogen. The angle $\alpha$ is therefore, equal to 70°, the tetrahedral structure of carbon valency being assumed.

$V_0$ is now the only parameter to be determined. This can be calculated from the experimental values of $m$, using Equations (1.2) and (1.3) and the validity of these equations can be shown by obtaining values of $V_0$ constant over the whole range of temperature.

Among the values of dipole moments of methyl halides determined by many authors, we shall use the following most probable ones — 1.85 $D$ for $CH_3Cl$,  1.80 $D$ for $CH_3Br$ and 1.65 $D$ for $CH_3I$.** Using these values of $\mu_x$ and $\mu_y$ and putting $\alpha = 70°$ as above, we can obtain a series of constant values of $V_0$ for each observed values of $m$ in a definite solvent. In Table 1.2 are collected the mean values of $V_0$ thus obtained.

TABLE 1.2

The values of $V_0 \times 10^{14}$ (in ergs) in various solvents

| Solvent | $\varepsilon$ | $ClH_2C$–$CH_2Cl$ | $ClH_2C$–$CH_2Br$ | $BrH_2C$–$CH_2Br$ | $IH_2C$–$CH_2I$ |
|---|---|---|---|---|---|
| Vacuum | 1.00 | 11.12 | 13.37 | 16.85 | — |
| Hexane | 1.90 | 8.13 | 10.83 | — | 41 |
| Heptane | 1.93 | 7.83 | 10.82 | 14.80 | — |
| Amylene | 2.06 | 7.64 | — | — | — |
| Ethyl ether | 4.37 | 6.65 | — | — | — |

We shall show in Table 1.3 how the calculated moments $(m_{calc})$ obtained by substituting these values of $V_0$ (of Table 1.2) into Equations (1.2) and (1.3) coincide with those observed $(m_{obs})$.

---

* The limiting value of the moment $m$ of the molecule, when $V_0 = 0$ or when there is complete freedom of rotation is evidently

$$m = \sqrt{2}\,\mu_0 \sin \alpha$$

which is identical with the equation derived by Williams for the case of two freely rotating dipoles. See J. W. Williams, Z. physik. Chemie 138, 75 (1928).

** $D$ is the Debye unit equal to $10^{-18}$ e. s. u.

TABLE 1.3

Comparison of observed and calculated moments*

(a)  $ClH_2C \cdot CH_2Cl$

| Gaseous state | | | Hexane solution | | | Heptane solution | | |
|---|---|---|---|---|---|---|---|---|
| $T$ | $m_{obs}$ $(D)$ | $m_{calc}$ $(D)$ | $T$ | $m_{obs}$ $(D)$ | $m_{calc}$ $(D)$ | $T$ | $m_{obs}$ $(D)$ | $m_{calc}$ $(D)$ |
| 304.95 | 1.12 | 1.15 | 223 | 1.13 | 1.15 | 223 | 1.16 | 1.17 |
| 341.03 | 1.24 | 1.23 | 248 | 1.21 | 1.22 | 243 | 1.24 | 1.23 |
| 376.25 | 1.32 | 1.30 | 273 | 1.30 | 1.29 | 263 | 1.31 | 1.29 |
| 419.00 | 1.40 | 1.38 | 298 | 1.36 | 1.36 | 283 | 1.36 | 1.34 |
| 456.96 | 1.45 | 1.44 | 323 | 1.42 | 1.42 | 303 | 1.41 | 1.39 |
| 479.82 | 1.48 | 1.47 | | | | 323 | 1.42 | 1.44 |
| 484.82 | 1.48 | 1.48 | | | | | | |
| 543.66 | 1.54 | 1.57 | | | | | | |

| Amylene solution | | | Ethyl ether solution | | |
|---|---|---|---|---|---|
| $T$ | $m_{obs}$ $(D)$ | $m_{calc}$ $(D)$ | $T$ | $m_{obs}$ $(D)$ | $m_{calc}$ $(D)$ |
| 223 | 1.22 | 1.20 | 213 | 1.25 | 1.26 |
| 248 | 1.29 | 1.28 | 223 | 1.29 | 1.29 |
| 273 | 1.33 | 1.34 | 233 | 1.32 | 1.33 |
| 293 | 1.38 | 1.39 | 243 | 1.35 | 1.35 |
| | | | 253 | 1.39 | 1.39 |
| | | | 263 | 1.43 | 1.42 |
| | | | 273 | 1.45 | 1.44 |
| | | | 283 | 1.48 | 1.47 |
| | | | 293 | 1.51 | 1.49 |

(b)  $ClH_2C \cdot CH_2Br$

| Gaseous state | | | Hexane solution | | | Heptane solution | | |
|---|---|---|---|---|---|---|---|---|
| $T$ | $m_{obs}$ $(D)$ | $m_{calc}$ $(D)$ | $T$ | $m_{obs}$ $(D)$ | $m_{calc}$ $(D)$ | $T$ | $m_{obs}$ $(D)$ | $m_{calc}$ $(D)$ |
| 338.93 | 1.08 | 1.08 | 248 | 0.95 | 1.03 | 223 | 0.93 | 0.96 |
| 368.22 | 1.12 | 1.14 | 273 | 1.04 | 1.08 | 243 | 0.99 | 1.01 |
| 405.28 | 1.19 | 1.20 | 298 | 1.14 | 1.14 | 263 | 1.06 | 1.05 |
| 435.60 | 1.27 | 1.25 | 323 | 1.21 | 1.19 | 283 | 1.12 | 1.10 |
| | | | | | | 303 | 1.21 | 1.15 |
| | | | | | | 323 | 1.20 | 1.19 |
| | | | | | | 343 | 1.20 | 1.23 |

* Most of the experimental values are taken from the measurement of Mizushima, Morino and Higasi, Sci. Pap. Inst. Phys. Chem. Res. Tokyo, **25**, 159 (1934); Physik. Z. **35**, 905 (1934). The moments observed by other authors (Smyth, Zahn, etc.) are recalculated by using the following values of $P_E + P_A$: (23.9 for $C_2H_4Cl_2$, 26.9 for $C_2H_4ClBr$, 29.9 for $C_2H_4Br_2$ and 39.9 c. c. for $C_2H_4I_2$).

(c)   $BrH_2C \cdot H_2CBr$

| Gaseous state | | | Hexane solution | | | Heptane solution | | |
|---|---|---|---|---|---|---|---|---|
| $T$ | $m_{obs}$ (D) | $m_{calc}$ (D) | $T$ | $m_{obs}$ (D) | $m_{calc}$ (D) | $T$ | $m_{obs}$ (D) | $m_{calc}$ (D) |
| 339.12 | 0.93 | 0.93 | 273 | 0.79 | 0.86 | 243 | 0.75 | 0.84 |
| 367.78 | 0.98 | 0.98 | 298 | 0.91 | 0.90 | 263 | 0.88 | 0.87 |
| 405.22 | 1.03 | 1.04 | 323 | 0.92 | 0.94 | 283 | 0.91 | 0.91 |
| 435.76 | 1.09 | 1.08 | | | | 303 | 0.97 | 0.94 |
| | | | | | | 323 | 0.97 | 0.98 |

(d)   $IH_2C \cdot CH_2I$  in  hexane  solution

| $T$ | $m_{obs}$ (D) | $m_{calc}$ (D) |
|---|---|---|
| 298 | 0.44 | 0.49 |
| 323 | 0.55 | 0.51 |

We are now interested in an explanation of the values of $V_0$ based on some molecular model.  Smyth, Dornte and Wilson[6] calculated the electrostatic interaction between bond dipoles on the two movable  groups of 1,2-dichloroethane and obtained  the  value  of  $2\,V_0$  as  $7 \sim 6 \times 10^{-14}$  ergs which is of the same order of magnitude as the experimental values shown in Table 1.2.  Moreover, if the electrostatic force were to play the predominant part, we would be justified in using the simple potential function of Eq. (1.1), because the internal potential would mainly be determined by the interaction between the two C–Cl bond dipoles which are much greater. than those of C–H.

However, as shown in Table 1.2 the value of $V_0$ for $BrH_2C–CH_2Br$ is much greater than that for $ClH_2C–CH_2Cl$ and this large difference cannot be explained by the electrostatic consideration, because the bond moment of C–Br is in fact less than that of C–Cl.  We must, therefore, consider that some force other than the electrostatic one must play a part in hindering the internal rotation.

We shall show in Section 12 that this force is the steric repulsion between the two rotating groups.  This is the reason why $V_0$ of $BrH_2C–CH_2Br$ is much greater than that of $ClH_2C–CH_2Cl$, as may easily be understood from the difference in the van der Waals radius of Br and Cl atoms.  If now the steric repulsion plays an important role in the hindering potential of internal rotation, we must consider the interaction between the hydrogen and the halogen atoms on different groups which we have so far neglected in our discussion of the hindering potential. If this interaction is fairly large, we should have two other potential minima in addition to the *trans* minimum. Accordingly, there is a distribution of molecules in these three potential troughs and the change of this distribution with temperature may account for the observed temperature dependence of dipole moment of the 1,2-dihalogenoethanes. This is precisely the picture of isomers whose structures can pass from one to the other by means of the internal rotation such as we have stated in the beginning of this chapter, but we cannot isolate any of these isomers, since the barrier separating the potential minima is so low that the equilibrium between the rotational isomers of 1,2-dihalogeno-ethanes is readily reached. We shall show in the following sections that this view of the coexistence of the rotational isomers or of the existence of several potential minima is preferable to the view of considering only one potential minimum as we have done in this section. It will be shown that the height of the potential barrier in such a simple molecule as halogeno-ethanes is less than 10 kcal./mole which is sufficiently low to account for the non-occurrence of stable rotational isomers.

Almost at the same time as the dipole investigations explained above, some diffraction investigations were made on the 1,2-dihalogenoethanes in the gaseous state. Wierl[10] who made the electron diffraction investigation tried to explain his experimental results by assuming the mixtures of the *trans* and the *cis* forms. He came to the conclusion that his results could not distinguish between the two possible cases of the *trans-cis* mixture and the predominant *trans* form for 1,2-dichloroethane, but the former possibility was ruled out for 1,2-dibromoethane. We cannot agree with the view of the existence of the stable *cis*-form, as we shall explain in detail in the next chapter, but Wierl's conclusion concerning the relation between 1,2-dichloroethane and 1,2-dibromoethane is in qualitative agreement with the results of the dielectric measurements, because the value of $V_0$ of the bromide is larger than that of the chloride: in other words, the tendency to assume the *trans* form is larger for the bromide than that for the chloride.

Ehrhardt[11] made X-ray diffraction investigations of gaseous 1,2-dichloro-ethane and discussed in detail the diffraction pattern from various angles. He could not support the view of the *trans-cis* mixture proposed by Wierl.[10] His results seemed to be explained by the same potential function (1.1) as that used in explaining the dielectric data.

---

10. R. Wierl, Ann. Physik, **13**, 453 (1932).
11. F. Ehrhardt, Physik, Z.,**33**, 605 (1932).

We shall show in the next chapter that more exact measurement of electron diffraction gives us a more detailed picture of the internal rotation of 1,2-dihalogenoethanes, but even the earlier experiments in this field gave the definite result that the internal rotation is never free, and that it is hindered to a considerable extent.

Table 1.2 also shows that the value of $V_0$ depends considerably on the solvent and that it decreases as the dielectric constant of the solvent becomes larger.*  We know many instances ·in which there is an intimate relation between the intramolecular energy and the dielectric constant of the medium (e. g. electrolytic dissociation), and that the energy decreases in a medium of higher dielectric constant as we have just seen for $V_0$ of 1,2-dihalogeno-ethanes.  However, we need more information about the internal potential, before we discuss the solvent effect in more detail.  (See Section 7.)

### 3. Some Other Earlier Dipole Investigations

Of interest in this connection is the problem of the optically active and *meso* forms of compounds of the tartaric acid type, $x\,y\,z\,C - C\,x\,y\,z$ (see Fig. 1.5).  For completely free rotation the two forms should have the same

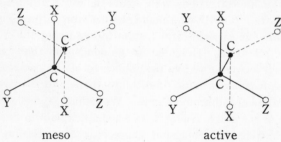

meso                                   active

Fig. 1.5.  Stable configurations of meso and active forms.

moment, so that any difference in the moment would indicate some restraint: three such pairs are shown in Table 1.4: hydrobenzoin and isohydrobenzoin[12,13] diethyl *dextro*- and *meso*-tartrate[14], and the two forms of stilbene dichloride[15].

---

* Benzene and toluene which are not included in the table are the exceptions.  We shall see in Section 7 that these two substances also behave abnormally in the relative intensity of Raman lines.

12. A. Weissberger and R. Sängewald, Z. phys. Chem. **B 12**, 399 (1931).
13. O. Hassel and E. Naeshagen, Z. phys. Chem. **B 14**, 232 (1931).
14. K. L. Wolf, Trans. Farad. Soc., **26**, 315 (1930).
15. A. Weissberger and R. Sängewald, Z. phys. Chem. **B 9**, 133 (1930).

TABLE 1.4

| Substance | | $\mu$ |
|---|---|---|
| Hydrobenzoin, $C_6H_5 \cdot HOHC–CHOH \cdot C_6H_5$ (meso) | | 2.08 |
| Isohydrobenzoin, (DL) | | 2.67 |
| Diethyl dextro-tartrate, | $12°$ | 3.09 |
| $C_2H_5OOC \cdot OH \cdot HC–CH \cdot OH \cdot COOC_2H_5$ | $38°$ | 3.16 |
| Diethyl meso-tartrate | $12°$ | 3.66 |
| | $38°$ | 3.69 |
| $\alpha$-Stilbene dichloride, $C_6H_5 \cdot Cl \cdot HC–CH \cdot ClC_6H_5$ | | 1.27 |
| $\beta$-Stilbene dichloride | | 2.75 |

If, with the more symmetrical meso form, two identical atoms or groups are in the trans position, — which would give no dipole moment, — the other two pairs of atoms or groups must also be in the trans position (see Fig. 1.5). With the optically active form, however, no entirely symmetrical configuration of the molecule can be obtained. The moments of stilbene chlorides can, therefore, be interpreted as indicating that the lower value corresponds to the meso and the higher values to the active form.

We see, however, that the difference between the moments of stereoisomers is smaller with hydrobenzoins than with stilbene dichlorides and that the order is reversed with the diethyl tartrates. The latter two cases seem to be due to the presence of internal hydrogen bonds by which the molecules tend to assume configurations with higher dipole moment, as will be explained in detail in Section 11. It may be generally concluded that the meso form has a moment value less than that of the active form except for some special cases such as that of the diethyl tartrates with internal hydrogen bonds.*

Another interesting case of restriction of free rotation is to be found in compounds of the type $C\alpha_4$, where $\alpha$ is a non-axially-symmetric substituent of the type $OCH_3$, $CH_2OH$, $CH_2ONO_2$, $CH_2Cl$, etc. Some of the observed dipole moments for compounds of this type are shown in Table 1.5. If completely free rotation of such substituents about the four tetrahedral valencies of the central carbon atom were possible, then the moment of the molecule would not be zero but would have a definite value given by $\sqrt{4\mu_1{}^2} = 2\mu_1$ where $\mu_1$ is the component of the group moment perpendicular to the axis of rotation. However, since some of the compounds like

---

* The presence of more than one rotation axes would correspond to another special case. The internal rotation about the second or third axis may result in the higher moment value for the meso form.

$C(CH_2Cl)_4$ have zero moment, and the moments of the other compounds do not change with temperature, it was concluded that, due to lack of space for rotation or to the high intramolecular fields, the groups in these molecules do not rotate freely but take up some definite or oriented positions (Estermann).[16]

TABLE 1.5

Dipole moments of molecules $C\alpha_4$

| $\alpha$ | $\mu$ | Reference |
|---|---|---|
| $O \cdot CH_3$ | 0.8 | Ebert et al.[17] |
| $O \cdot C_2H_5$ | 1.1 | Ebert et al.[17] |
| $CH_2OH$ | ca. 2 | Estermann[16] |
| $CH_2O \cdot CO \cdot CH_3$ | 2.6 | Ebert et al.,[17] Estermann[16] |
| $CH_2O \cdot NO_2$ | 2.0 | Ebert et al.[17] |
| $CO \cdot OCH_3$ | 2.8 | Ebert et al.[17] |
| $CH_2Cl$ | 0 | Ebert et al.[17] |
| $CH_2Br$ | 0 | Ebert et al.,[17] Williams,[18] Estermann[16] |
| $CH_2I$ | 0 | Ebert et al.[17] |

There was much discussion at one time about the structure of molecules of this type, especially penterythritol $C(CH_2OH)_4$ and its derivatives. Some authors suggested, on the grounds of X-ray and spectroscopic measurements, that a molecule $C\alpha_4$ could have the form of a square pyramid with the carbon at the apex.[19] This caused a great sensation at that time, since the tetrahedral valency of the carbon atom has been the most fundamental concept in classical stereo-chemistry. If such a suggestion were accepted, all the dipole moment data would have to be explained quite differently. Later work, however, threw great doubt on the conclusions of square pyramidal structure: for example, Nitta[20] and others showed that the conclusion based on crystal structure was due to a misinterpretation of experimental results. At present we have no spectroscopic, nor X-ray evidence according to which the classical concept of tetrahedral carbon can be called into question.

16.  J. Estermann, Leipziger Vorträge 1929, 17.
17.  L. Ebert, R. Eisenschitz and H. von Hartel, Z. phys. Chem. B 1, 94 (1928).
18.  J. W. Williams, Physik. Z. 29, 683 (1928).
19.  H. Mark and K. Weissenberg, Z. Physik, 17, 301 (1926).
20.  I. Nitta, Bull. Chem. Soc. Japan, 1, 62 (1926).

## Summary of Chapter I.

The internal rotation about a carbon single bond as axis was for some time supposed to be entirely free and even the term "free rotation" has often been used in chemistry. However, considerable evidence based on the various physical methods has been accumulated to indicate that the principle of free rotation is not correct.

In this chapter some dipole investigations have been explained as earlier studies on this problem. If there is completely free rotation about the carbon-carbon single bond of such a molecule as 1,2-dichloroethane, $ClH_2C-CH_2Cl$, the dipole moment of the molecule would be independent of temperature. This would also be the case, if there is only one sharp potential minimum of internal rotation. Actually the value of the dipole moment of 1,2-dichloro-ethane was found to change considerably with temperature and accordingly we can exclude both the free rotation, and the existence of one sharp minimum.

The observed temperature dependence of dipole moment can be explained by assuming a fairly broad potential minimum about which the two movable groups exert torsional vibrations of large amplitude, or by assuming more than one potential minimum of different depths. The result of the calculation based on the first assumption could satisfactorily explain the observed change of dipole moment with temperature. However, this experimental result can be explained by the second assumption equally well and the dipole data alone cannot decide which of the two assumptions is preferable.

# CHAPTER II

# Ethane and Its Derivatives I
## (Spectroscopic and Electric Measurements)

### 4. The Raman Effect in Ethane Derivatives

A molecule containing $N$-atoms has $3N - 6$ normal vibrations (in the case of a linear molecule $3N - 5$) and, therefore, even if all of them are allowed in the Raman effect by the selection rule, there cannot in general be observed more than $3N - 6$ Raman lines (fundamental frequencies). The overlapping by other lines (the accidental coincidence of some frequencies) may further reduce the apparent number of Raman lines.

It is, therefore, remarkable to find the number of lines much larger than this value for molecules having internal rotation axes and we have to conclude that such molecules should have more than one configuration.* The molecules of 1,2-dihalogenoethanes in the liquid and gaseous states belong to such a category as will be seen from the experimental results obtained by Mizushima, Morino et al.[1, 2, 3, 4, 5, 6] shown in Fig. 2.1.

In the solid state, however, the Raman lines are materially reduced in number, as shown in the same figure, and the spectra can be explained by considering only one molecular configuration. It seems, therefore, advantageous to start the study of internal rotation by determining the molecular configuration in the solid state.

---

* See, for example: G. Glockler and C. Sage, J. Chem. Phys., 9, 387 (1941).

---

1. S. Mizushima, Y. Morino and K. Higasi, Sci. Pap. Inst. Phys. Chem. Res. Tokyo, 25, 159 (1934).
2. S. Mizushima, Y. Morino and S. Noziri, Sci. Pap. Inst. Phys. Chem. Res. Tokyo, 29, 63 (1936).
3. S. Mizushima and Y. Morino, Proc. Ind. Acad. Sci. 8, 315 (1938). Raman Jubilee Volume.
4. S. Mizushima and Y. Morino, Bull. Chem. Soc. Japan, 17, 94 (1942).
5. S. Mizushima, Y. Morino, Y. Miyahara, M. Tomura and Y. Okamoto, Sci. Pap. Inst. Phys. Chem. Res. Tokyo, 39, 387 (1942).
6. Y. Morino, I. Watanabe and S. Mizushima, Sci. Pap. Inst. Phys. Chem. Res. Tokyo, 39, 396 (1942).

From a comparison of the solid spectra of Fig. 2.1 among themselves, it is readily seen that the number of Raman lines of the molecules with two like halogen atoms is smaller than that of the molecules with two different halogen atoms. For example, in the region of the carbon-halogen stretching frequencies* $(500 \sim 800 \text{ cm}^{-1})$ there have been observed two Raman lines

Fig. 2.1  Raman spectra of 1,2-dihalogenoethanes in the liquid and solid states.

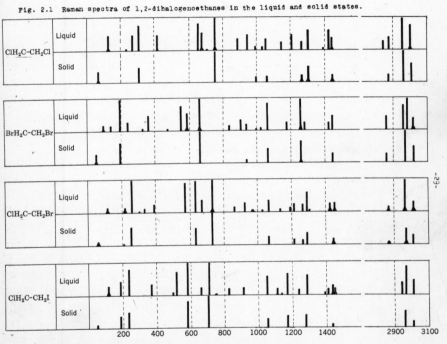

Fig. 2.1.  Raman spectra of 1.2-dihalogenoethanes in the liquid and solid states.

for the latter, but only one for the former. Therefore, the stable configuration of the molecule with two like halogen atoms in the solid state can be considered to have a center of symmetry, since in this case only the symmetric vibration is allowed in the Raman effect. We call this configuration the *trans* form which is a staggered form with two halogen atoms as far apart as possible (see Fig. 2.2 *a*). The Raman-active normal frequencies calculated

---

* For the meaning of such a terminology as C–Cl stretching frequency, see Section 9, Part II.

for this form can be shown to be in good agreement with those actually observed in the Raman effect.*   Therefore, we have to conclude that there is only one significant potential minimum in one complete rotation about a C–C single bond as axis and all the molecules in the solid state tend to assume the *trans* form corresponding to this minimum.  All the configurations corresponding to other positions of internal rotation are unstable or far less stable than the *trans* form.

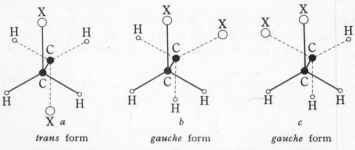

|  |  |  |
| --- | --- | --- |
| *trans* form | *gauche* form | *gauche* form |

Fig. 2.2.   Stable configurations of 1,2-dihalogenoethanes.

Now since all the spectral lines observed in the solid state remain almost unchanged in frequency in the liquid state, it is concluded that the *trans* form is a stable configuration in the liquid state, although there may be another configuration giving rise to the Raman lines which are observed in the liquid state only.  As to these lines, they may be assigned to vibrations of the *trans* form which become active owing to the distortion of molecules by the more irregular interaction between molecules in the liquid state. Mizushima, Morino and Shimanouchi[7] have shown by the more careful measurement of the Raman effect that such is actually the case for a few lines observed only in the liquid state.  For example, the weak Raman lines of 1,2-dichloroethane observed at 223 cm$^{-1}$ and at 709 cm$^{-1}$ should be forbidden by the selection rule, since both of them are assigned  to the vibrations antisymmetric to the center of symmetry of the *trans* form.** They are considered to become active because of the distortion of molecules in the liquid state.  However, we can show by the infrared measurement,

---

* For the calculation of normal vibrations of 1,2-dihalogenoethanes, see Section 6 and 9, Part II.

** See Section 2, Part II.

---

7.  S. Mizushima, Y. Morino and T. Shimanouchi, Sci. Pap. Inst. Phys. Chem. Res. Tokyo, **40**, 87 (1942).

to be explained in the next section, that most of the lines observed only in the liquid state should be assigned to normal vibrations of another configuration.

So far as the Raman effect is concerned, the simplest proof of the existence of the second stable form and the simplest method for the determination of this form would be provided by the measurement of tetra- and penta-substituted ethanes, e. g. $ClH_2C–CCl_3$, $Cl_2HC–CCl_3$ etc. According to the Raman measurements made by Mizushima, Morino and others[8,9] for these substances, there has been observed no difference between the spectra of the liquid and solid states and the number of the observed lines is small enough to be explained on the basis of only one molecular configuration.

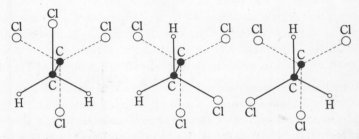

Fig. 2.3.   Stable configurations of 1,1,1,2-tetrachloroethane.

We can see from what we shall discuss in Section 8 that the potential minima of the internal rotation of such molecules would correspond to the three configurations as shown in Fig. 2.3. These configurations are identical and therefore, if the internal state of such a molecule is changed from any one of them to another, no change in the Raman spectrum will be observed. Experimentally no change in the Raman spectrum is observed.[8]

It is very probable then that there are similar potential minima in one complete internal rotation of 1,2-dihalogenoethanes, the corresponding molecular configurations being shown in Fig. 2.2 and that we can assign the Raman lines observed only in the liquid state to the configurations b and c of Fig. 2.2, which are the other staggered varieties obtained by rotating one end of the *trans* form through an angle of $\pm$ 120°. Let us call

8.  S. Mizushima, Y. Morino, M. Kawano and R. Ochiai, Sci. Pap. Inst. Phys. Chem. Res. Tokyo, 42, 1 (1944).

9.  Y. Morino, S. Yamaguchi and S. Mizushima, Sci. Pap. Inst. Phys. Chem. Res. Tokyo, 42, 5 (1944).

these two configurations the *gauche* forms.   They are the mirror images of
each other and will not be different in their normal frequencies.

There are other reasons to select the *gauche* form as the second stable
form.  As will be shown in Section 8, Part II, a deformation frequency of
the heavy atom skeleton of the dihalogenoethane depends considerably on
the azimuthal angle of internal rotation.  In other words, from the observed
value of the deformation frequency we can tell fairly accurately the azimuthal
angle of the *gauche* form, which is found to be about 120°.  The same result
is obtained by the application of the product rule for rotational isomers
derived by Mizushima, Morino and Shimanouchi.[7]  According to this rule
the ratio of the product of the vibrational frequencies of one rotational
isomer to that of the other isomer can be calculated from the molecular
configuration which, in this case, depends only on the azimuthal angle of
internal rotation.  Therefore, we can determine the azimuthal angle if we
know all the vibrational frequencies of both of the two rotational isomers.
We shall discuss this problem in more detail in Section 8 of Part II.

Based on the experimental data of De Hemptinne on deuterium-
substituted dibromoethanes[10] Neu and Gwinn[11] arrived at the same conclusion for the second stable form.  The essential feature of their argument
is as follows  (see Table 2.1).  If this form were the *cis* form as was suggested
by some authors, the four hydrogen atoms in this form of $BrH_2C-CH_2Br$
would be equivalent, and the molecules $BrDHC-CH_2Br$ and $BrDHC-CD_2Br$
would each give only one second stable configuration and would allow the
same number of Raman lines as do $BrH_2C-CH_2Br$ and $BrH_2C-CD_2Br$.  If,
on the other hand, the second isomer is in the *gauche* form, two different
spatial configurations would be possible for this isomer of both $BrDHC-CH_2Br$
and $BrDHC-CD_2Br$ (see Table 2.1) and two separate but similar Raman
spectra originating from the two configurations would be expected to appear.
$BrH_2C-CH_2Br$, $BrH_2C-CD_2Br$ and $BrD_2C-CD_2Br$ in the *gauche* form would
exist in only one spatial configuration.

In Table 2.1 are shown some of the low frequency Raman lines of the
isotopic molecules of 1,2-dibromoethane.  We see from this table that as
deuterium atoms are added, the frequencies assigned to the second stable
form split and converge exactly as expected for the *gauche* form.  (Each

---

10. M. de Hemptinne, Contrib. etude structure mol. Vol. commem. Victore Henri,
151 (1947/48).

11. J. T. Neu and W. D. Gwinn, J. Chem. Phys. **18**, 1642 (1950).

successive substitution of a deuterium atom lowers the frequencies slightly.) Thus we have further evidence that the second stable form is the *gauche* form* but not the *cis* form.

TABLE 2.1

Molecular configurations and some Raman frequencies of 1,2-dibromoethane **

| Form | $BrH_2C–CH_2Br$ | $BrH_2C–CHDBr$ | $BrH_2C–CD_2Br$ | $BrHDC–CD_2Br$ | $BrD_2C–CD_2Br$ |
|---|---|---|---|---|---|
| Trans | *(structure)* | *(structure)* | *(structure)* | *(structure)* | *(structure)* |
| Cis | *(structure)* | *(structure)* | *(structure)* | *(structure)* | *(structure)* |
| Gauche | *(structure)* | *(structures)* | *(structure)* | *(structures)* | *(structure)* |
| Trans | 188(10) | 188(10) | 188(5) | 187(10) | 188(10) |
| Gauche | 355(2) | 342(1) / 352(1) | 341(1) | 321(1) / 333(1) | 316.5(2) |
| Gauche | 552(5) | 532(4) / 546(4) | 528(3) | 518.5(4) / 523(2) | 517.5(6) |
| Gauche | 581(1) | 565.5(2) / 580(3) | 551(1) | 540(2) | 526.5(3) |
| Trans | 660(15) | 640(15) | 639(6) | 617(10) | 607(15) |

--------

* The azimuthal angle of the *gauche* form would not be exactly 120°, but would not be much different from this value. We shall discuss this problem again in Section 8 of Part I and in Section 8 of Part II.

** This table was taken from the paper of Neu and Gwinn. See ref. 11.

The difference between the Raman spectra in the liquid and solid states must be due to the difference in the molecular interaction in these two states. In this respect the measurement of the Raman spectrum in the gaseous state would be very interesting, since in this state the interactions between molecules would become negligible. Such measurements were actually made by Morino, Watanabe and Mizushima[6, 12] and it was shown that, so far as the frequencies of lines are concerned, there is no fundamental difference between the spectra in the liquid and gaseous state. This shows that the free molecules of dihalogenoethanes are also in the *trans* and *gauche* forms. However, the relative intensity of Raman lines observed in the gaseous state is different from that in the liquid state and we have to conclude that the equilibrium ratio of the rotational isomers in the gaseous state is different from that in the liquid state. The numerical values of this ratio will be discussed in Section 6 after we have explained the infrared and dielectric data.

Thus we have seen that the Raman spectra of 1,2-dihalogenoethanes can never be explained by such a simple potential curve as shown in Fig. 1.3. The spectroscopic results require the existence of a potential minimum or minima other than the *trans* one. This is just the picture of the coexistence of rotational isomers whose structures can pass into one another by means of the internal rotation about the carbon-carbon single bond as axis. This picture was one of the possibilities for the explanation of the temperature-dependence of the dipole moment, so far as we confine ourselves to the discussion on the experimental results of dipole measurement from which we know only the mean state of the molecules. However, by the measurement of the Raman effect in which different kinds of molecules give different patterns we can conclude that the view of the coexistence of rotational isomers is preferable to the view of the existence of one broad minimum which was the other possible explanation of the results of the earlier dipole investigations. Originally the existence of more than one potential minimum, or the idea of rotational isomers, was proposed by Kohlrausch who, however, assigned the stable configurations as the *trans* and the *cis* forms, just as in the case of ethylene derivatives.[13] Later work showed that the *cis* form cannot be stable and the idea of the *gauche* form was proposed in 1937 by Mizushima, Morino and Kubo.[14] The existence of this form was proved

12. S. Mizushima, Y. Morino, I. Watanabe, T. Shimanouchi and S. Yamaguchi, J. Chem. Phys. 17, 591 (1949).

13. K. W. F. Kohlrausch, Z. physik. Chem. B 18, 61 (1932).

14. S. Mizushima, Y. Morino and M. Kubo, Physik. Z., 38,. 459 (1937).

experimentally in 1941 by Edgell and Glockler[15] and Mizushima, Morino and Takeda[16] and later on by many other investigators.

The experimental technique of observing the spectral difference between the liquid and solid states provides, as stated above, the simplest method of detecting rotational isomerism in the liquid state. This technique of the investigation of rotational isomerism was first applied to 1,2-dichloroethane by Mizushima, Morino and Noziri[2] in 1936 and has been applied to many other substances by them as well as by other investigators. We shall see in Section 18 that the complexity of the spectra of long chain paraffin molecules is reduced by this technique to such an extent that the interpretation of the spectrum of at least one of the isomeric forms is possible.

## 5. Infrared Absorption of Ethane Derivatives

Infrared absorption measurements can be applied to the internal rotation investigation in a manner quite similar to the application of the Raman effect. In Table 2.2 are shown the absorption spectra observed in the rock salt and sylvine regions for 1,2-dihalogenoethanes[12, 17, 18, 19, 20] together with the Raman spectra of the same substances.

Therefore, in the case of 1,2-dichloroethane and 1,2-dibromoethane, shown in Table 2.2, all the absorption bands, observed in the solid state, should be assigned to the normal vibrations of the *trans* molecule which are antisymmetric to the center of symmetry according to the requirement of the selection rule of infrared absorption. Now since only the symmetric vibrations are allowed in the Raman effect, none of the Raman frequencies should agree with the infrared absorption frequencies except for accidental coincidence (see Section 2, Part II). This was actually the case for the spectra observed in the solid state shown in Table 2.2 and, accordingly, both the Raman and infrared investigations support the view that practically all the molecules in the solid state are in the *trans* form.

15. W. F. Edgell and G. Glockler, J. Chem. Phys. 9, 375 (1941).

16. S. Mizushima, Y. Morino and M. Takeda, J. Chem. Phys. 9, 826 (1941); Sci. Pap. Inst. Phys. Chem. Res. Tokyo, 38, 437 (1941).

17. H. C. Cheng and J. Lecomte, J. Phys. Rad., 6, 477 (1935).

18. T. Shimanouchi, H. Tsuruta and S. Mizushima, Sci. Pap. Inst. Phys. Chem. Res. Tokyo, 42, 51 (1944).

19. J. K. Brown and N. Sheppard, Faraday Soc. Discussion, 9, 144 (1950); Trans. Farad. Soc., 48, 128 (1952).

20. K. Kuratani, T. Miyazawa and S. Mizushima, unpublished.

TABLE 2.2

Infrared and Raman spectra of 1,2-dihalogenoethanes (observed by Mizushima, Morino, Shimanouchi, Kuratani, Miyazawa, et al.)*

(a) 1,2-dichloroethane

| Raman | | | Infrared | | |
|---|---|---|---|---|---|
| Gas | Liquid | Solid | Gas | Liquid | Solid |
| | | 66(4b) | | | |
| | 125(5b) | | | | |
| | 223(00) | | | | |
| | 265(5) | | | | |
| 301(7) | 300(8) | 302(5) | | | |
| | 411(5) | | | | |
| 666(4) | 654(8) | | | 655 (s) | |
| 689(0) | 677(6b) | | | 675 (s) | |
| | (709)(0) | | 725 (vs) | 709 (vs) | 710 (s) |
| 768(10) | 754(10b) | 750(10) | | 753(vw)? | 749 (vw) |
| | | | ca. 770 (w) | 768 (m) | 774 (m) |
| | | | | ca. 820(vw)? | |
| | 881(4) | | 889 (m) | 880 (vs) | |
| 950(0)? | 943(5) | | 945 (w) | 941 (vs) | |
| | | | | ca. 980(vw)? | 980 (vw) |
| | 989(2) | 992(3) | | | |
| | | | | 1008 (w) | 1004 (w) |
| | 1031(2) | | 1025 (w) | 1031 (m) | 1030 (vw) |
| 1040(0)? | 1052(4) | 1059(3) | | | |
| | | | 1123 (w) | 1124 (w) | 1127 (m) |
| | 1145(3) | | | 1140 (w) | |
| | 1207(5) | | | | |
| | | | 1232 (vs) | 1230 (vs) | 1233 (s) |
| | 1264(3) | 1265(2b) | | | |
| | | | 1291 (s) | 1284 (s) | |
| 1305(3) | 1304(6) | 1301(5b) | | | |
| | | | 1318 (sh) | 1312 (s) | |
| | 1393(1) | | | | |
| | 1429(6) | | 1435 (sh) | 1430 (vs) | |
| | 1445(4b) | 1448(2b) | 1450 (m) | 1450 (s) | 1452 (s) |
| | | | | ca. 1500 (vw) | 1472 (vw) |
| | 2844(3) | | | | |
| 2887(1)? | 2874(4) | 2874(3) | 2890 (w) | 2870 (w) | 2880 (w) |
| 2962(8) | 2957(10) | 2964(10) | 2975 (m) | 2960 (s) | 2975 (m) |
| 2978(8) | 3005(8b) | 3005(6) | | | 3040 (vw) |

* The number in parenthesis after each Raman frequency refers to the visually estimated intensity on the plate and accordingly is quite different from the actual relative intensity. The infrared spectra were measured in the frequency region between 500 and 3300 cm.$^{-1}$. See ref. 3, 5, 18 and 20.

(b)  1-chloro, 2-bromoethane

| Raman | | Infrared | | |
|---|---|---|---|---|
| Liquid | Solid | Gas | Liquid | Solid |
| | 57(1b) | | | |
| 107(2b) | | | | |
| 210(2b) | 210(0) | | | |
| 251(10) | 248(6) | | | |
| 300(0,0) | | | | |
| 327(1) | | | | |
| 385(3) | | | | |
| 568(9) | | | 569 (s) | |
| 630(9) | 629(6) | | 635 (m) | 630 (s) |
| 665(6) | | | 665 (m) | |
| 726(10b) | 721(10) | 742 (vs) | 727 (s) | 723 (s) |
| | | 764 (vw) | 759 (m) | 762 (m) |
| 855(2) | | 867 (m) | 855 (s) | |
| | | | 878 (vw) | 876 (w) |
| 919(3) | | 928 (w) | 920 (s) | 923 (w) |
| 961(1b) | | | ca.  960 (vw) | 959 (w) |
| 1023(1) | | ca. 1025 (vw) | 1025 (m) | |
| 1052(4) | 1056(3) | 1055 (w) | 1055 (m) | 1058 (m) |
| | | 1110 (w) | 1111 (m) | 1109 (m) |
| 1128(1) | | 1124 (w) | 1125 (m) | |
| 1189(2) | | | 1188 (sh) | |
| 1203(3) | 1203(2) | 1204 (vs) | 1202 (s) | 1204 (s) |
| 1259(3) | 1258(2) | 1268 (s) | 1259 (s) | 1258 (vw) |
| 1284(7) | 1282(4) | | 1285 (sh) | 1284 (m) |
| 1299(1) | | 1296 (w) | 1299 (s) | |
| | | 1311 (w) | 1323 (vw) | |
| 1421(3b) | | 1427 (sh) | 1422 (s) | |
| 1444(3b) | 1439(2b) | 1446 (m) | 1440 (s) | 1444 (s) |
| | | | ca. 1500 (vw) | 1495 (vw) |
| 2868(2b) | 2865(1b) | 2875 (w) | 2865 (m) | 2875 (w) |
| 2960(10b) | 2964(5b) | 2980 (s) | 2960 (s) | 2975 (m) |
| 3010(3b) | 3007(3) | | 3010 (sh) | 3020 (w) |

In the liquid and the gaseous states there appear, in addition to the same frequencies as observed in the solid state, other absorption frequencies which, as stated above, are assigned to the normal vibrations of the *gauche* form.  As this form has no center of symmetry, the normal frequencies of the *gauche* form are allowed in both the Raman effect and infrared absorption. We see from Table 2.2 that this is actually the case in the frequency region from 500 to 3300 cme$^{-1}$ in which both the Raman and infrared measurements have been made.

(c) 1,2-dibromoethane

| Raman | | Infrared | | |
|---|---|---|---|---|
| Liquid | Solid | Gas | Liquid | Solid |
| | 49(3b) | | | |
| 91(2b) | | | | |
| 132(2) | | | | |
| 190(10) | 190(7) | | | |
| 231(3) | | | | |
| 325(1) | | | | |
| 355(5) | | | | |
| 469(1) | | | | |
| 551(8) | | | 551 (m) | |
| 583(6b) | | | 587 (s) | 584 (s) |
| 660(10b) | 657(10) | | | |
| | | 742 (w) | 751 (s) | 748 (s) |
| | | 786 (w) | 772 (m) | 770 (w) |
| | | | | 791 (vw) |
| 836(2) | | 840 (m) | 833 (s) | |
| | | | 853(vw)? | 854 (w) |
| 899(3) | | 890 (w) | 897 (m) | |
| 933(2) | 934(1) | | 926 (w) | |
| 997(0) | | | | |
| 1019(1) | | ca. 1020 (vw) | 1016 (m) | |
| 1053(9) | 1056(5) | | 1057 (w) | 1059 (w) |
| | | 1095 (w) | 1088 (m) | 1094 (m) |
| | | | 1104 (w) | |
| 1169(3) | | | | |
| | | 1187 (vs) | 1188 (s) | 1189 (s) |
| | | | 1203 (sh) | |
| | | 1251 (s) | 1246 (s) | |
| 1255(10b) | 1250(8b) | | | 1254 (w) |
| 1276(3) | | 1285 (w) | 1277 (s) | |
| | | | 1380 (w) | 1379 (w) |
| | | | | 1399 (vw) |
| 1419(3) | | 1431 (m) | 1420 (s) | |
| 1440(5) | 1438(3) | 1444 (m) | 1437 (s) | 1433 (s) |
| 2859(4) | 2857(3) | 2870 (w) | 2860 (m) | 2860 (vw) |
| 2953(8) | | | | |
| 2972(10) | 2966(10) | 2970 (m) | 2960 (s) | 2970 (m) |
| 3013(4b) | 3013(5) | | 3012 (m) | 3040 (m) |

In the case of 1-chloro, 2-bromoethane, $ClH_2C-CH_2Br$,[19, 20, 21] we also observe many more Raman lines or absorption peaks in the liquid state than in the solid state. This is due to the fact that this substance exists in two molecular forms in the liquid state and in only one form in the solid state. However, even in the *trans* form, in which all the solid molecules exist, there is no center of symmetry and, therefore, the Raman frequencies observed in the solid state can coincide with the infrared frequencies in contradistinction to the case of 1,2-dichloroethane or 1,2-dibromoethane. We see from Table 2.2 that such is actually the case.

The Raman and infrared spectra of 1,1,2,2-tetrachloroethane, $Cl_2HC-CHCl_2$ and 1,1,2,2-tetrabromoethane have been observed by many investigators.[22, 23, 24, 25, 26, 27] From the experimental results it can be concluded that the molecules exist in two different staggered forms (the *trans* and the *gauche* forms shown in Fig. 2.4) in the liquid state and in only one in the solid state, just as in the case of 1,2-dihalogenoethanes. Some authors[23] prefer an eclipsed form to the *trans* form, one of the staggered forms,* but the electron diffraction experiment by Schomaker and Stevenson[28] definitely rules out the eclipsed form.

We have seen in the preceding section that such a kind of molecule as $ClH_2C-CCl_3$, $Cl_2HC-CCl_3$, etc. exists only in one form (staggered form) in both the liquid and solid states. It would then be very probable that the molecule of hexachloroethane $Cl_3C-CCl_3$ also exists in one form which is the staggered form with $D_{3d}$ symmetry. This can be shown to be actually the case from the consideration of the selection rules for the Raman effect and infrared absorption.

---

* Herzberg questioned the assumptions which were used by these authors to arrive at the conclusion for the stability of the eclipsed form. See G. Herzberg, "Infrared and Raman Spectra of Polyatomic Molecules, D. van Nostrand Co., New York (1945).

---

21. S. Mizushima, Y. Morino, Y. Miyahara, M. Tomura and Y. Okamoto, Sci. Pap. Inst. Phys. Chem. Res. Tokyo, **39**, 387 (1942).

22. S. Mizushima, Y. Morino and K. Kozima, Sci. Pap. Inst. Phys. Chem. Res. Tokyo, **29**, 111 (1936).

23. A. Langseth and H. J. Bernstein, J. Chem. Phys., **8**, 410 (1940).

24. N. Sheppard and G. J. Szasz, J. Chem. Phys., **18**, 145 (1950).

25. J. Powling and H. J. Bernstein, J. Am. Chem. Soc., **73**, 1815 (1951).

26. R. E. Kagarise and D. H. Rank, Trans. Farad. Soc., **48**, 394 (1952).

27. K. Naito, K. Kuratani, I. Ichishima and S. Mizushima, unpublished.

28. V. Schomaker and D. P. Stevenson, J. Chem. Phys., **8**, 637 (1940).

The Raman and infrared spectra of hexachloroethane have been measured by several authors* including Mizushima, Morino, Shimanouchi and Kuratani[29] with the results shown in Table 2.3. The normal vibration calculation of this molecule for the $D_{3d}$ model was made by Shimanouchi[30] and by use of the result we can assign the observed frequencies as shown in the same table, in which the notations have the following significance: $A_{1g}$ is a vibration symmetric with respect to the center(i), the axis $(C_3)$

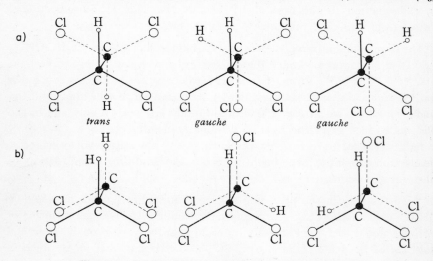

Fig. 2.4. Configurations of 1,1,2,2-tetrachloroethane.
(*a*) Staggered forms, (*b*) Eclipsed forms.

and the plane $(\sigma_v)$ of symmetry: in other words the totally symmetric vibration, $A_{2u}$ is a vibration symmetric with respect to the axis, but anti-symmetric with respect to the center, and $E$ is a doubly degenerate vibration which is differentiated by subscripts $g$ and $u$ according as the vibration is symmetric or anti-symmetric with respect to the center of symmetry.

All the frequencies expected to appear by the selection rules have been observed except the three infrared frequencies which lie outside the observable

---

* See, for example, S. Mizushima and Y. Morino, Sci. Pap. Inst. Phys. Chem. Res. Tokyo, **29**, 188 (1936); D. T. Hamilton and F. F. Cleveland, J. Chem. Phys. **12**, 249 (1944).

29. S. Mizushima, Y. Morino, T. Shimanouchi, and K. Kuratani, J. Chem. Phys. **17** 838 (1949).

30. T. Shimanouchi, J. Chem. Phys. **17**, 734 (1949).

TABLE 2.3

The assignment of vibrational frequencies of hexachloroethane.[29] (Figures in parentheses are the calculated frequencies)

| Class | Raman frequency | Class | Infrared frequency |
|-------|-----------------|-------|--------------------|
| $A_{1g}$ | 169<br>432<br>974 | $A_{2u}$ | (382)<br><br>675 |
| $E_g$ | 224<br>341<br>853 | $E_u$ | (134)<br>(278)<br>769 |

region of the rocksalt and sylvine monochromators. Especially important in determining the configuration of this molecule is the fact that the frequency, 769 cm$^{-1}$, belonging to $E_u$ appears in the infrared absorption and not in the Raman effect. This shows that the molecule is in the staggered form. If the molecule is in the eclipsed form, this frequency should appear in both of the infrared and Raman spectra. Moreover, there was found no coincidence of the Raman and infrared frequencies which is consistent with the existence of a center of symmetry. Thus we can conclude that the equilibrium configuration of this molecule is the staggered form.[29] We shall see below in Section 8 that this conclusion is compatible with the result of the electron diffraction experiments.

As we shall use hereafter such notations as referred to above to describe the symmetry property of a normal vibration, it would be appropiate to explain here the significance of such notations.

The configuration of a molecule or the geometric arrangement of nuclei possesses symmetry and there are a number of geometric operations, called covering operations, which can be performed on the molecule so that equivalent nuclei occupy the same points in space as in the original configuration. There are only two kinds of covering operations, viz. proper rotations and improper rotations. A proper rotation is simply a rotation through an angle $\theta$ about some axis of symmetry. An improper rotation is a rotation followed by a reflection in a plane perpendicular to the axis of rotation, therefore, a reflection is an improper rotation through an angle of 0°.

For hexachloroethane the principal symmetry axis $Z$ (always considered to be vertical) is the $C - C$ axis of internal rotation. The covering operations for this configuration are: (1) the identity operation $E$, which is a rotation through 0° about the $Z$ axis. (2) the operation $C_3$ which means a rotation through an angle $\pm 120°$ about the $Z$ axis. (3) the operation i, an inversion which is an improper rotation through an angle of 180° about the $Z$ axis and (4) the operation $\sigma_v$ which means a reflection in a plane passing through the $Z$ axis and one of Cl atoms.* Thus there is a set of four operations which constitutes a point group designated by the symbol $D_{3d}$. The point group for the *trans* form of 1,2-dichloroethane is $C_{2h}$, which contains a reflection in a horizontal plane passing through the $C - C$ axis and two Cl atoms and rotation through an angle of 180° about the $Z$ axis perpendicular to the $C - C$ axis. The point group for the *gauche* form is $C_2$ which contains a rotation through 180° about the $Z$ axis (Fig. 2.5).

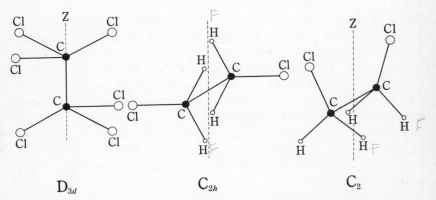

Fig. 2.5. The point groups for the stable configurations of hexachloroethane and 1,2-dichloroethane.

Since the potential energy of a molecule must have the same symmetry as the equilibrium configuration of the molecule, the symmetry property of a normal vibration is closely related to that of the molecule, as we have just seen in the case of hexachloroethane. In general the symmetry property of a type of normal vibration is designated by the following notations:

$A$ and $B$ are non-degenerate types and are, respectively, symmetric and antisymmetric to rotation about the $Z$ axis.

$E$ and $T$ are, respectively, doubly and triply degenerate types of vibration.

---

* The operation of a reflection in a plane is designated by a notation $\sigma$, to which subscripts $v$, $h$ and $d$ are added, according as the plane is vertical, horizontal and diagonal.

Types that are symmetric or antisymmetric to $\sigma_h$ are designated by ' or '', respectively (except in the case D'$_{3h}$).

Vibrations that are symmetric or antisymmetric to a center of symmetry are designated by the subscripts $g$ and $u$. The subscripts 1 and 2 indicate that the vibration is symmetric or antisymmetric with some of the other classes of the group not mentioned above.

If we classify the normal frequencies of the *trans* and the *gauche* forms of 1,2-dichloroethane by use of these notations, the observed frequencies can be assigned as shown in Table 2.4.*

TABLE 2.4

The assignment of vibrational frequencies of 1,2-dichloroethane

| Class | Activity | Observed frequencies | Class | Activity | Observed frequencies |
|---|---|---|---|---|---|
| $A_g$ | Raman active<br><br>Infrared inactive | 300<br>754<br>1052<br>1304<br>1445<br>2957 | $A$ | Raman active<br><br>Infrared active | 265<br>654<br>1031<br>1264<br>1429<br>2957 |
| $A_u$ | Raman inactive<br><br>Infrared active | —<br>768<br>1124<br>3005 | | | 125<br>943<br>1207<br>3005 |
| $B_g$ | Raman active<br><br>Infrared inactive | 989<br>1264<br>3005 | $B$ | Raman active<br><br>Infrared active | 881<br>1145<br>3005 |
| $B_u$ | Raman inactive<br><br>Infrared active | 223<br>709<br>1230<br>1450<br>2957 | | | 411<br>677<br>1304<br>1429<br>2957 |

* As to the procedure of assignments, see Section 9, Part II.

## 6. Dielectric Constant of Halogenoethanes

Now that we know that the molecules of 1,2-dihalogenoethanes in the gaseous state exist in the *trans* and the *gauche* forms, we have to reinterpret the experimental results of the dipole measurements which have been treated in Section 2 by the simple potential curve shown in Fig. 1.3.

From the point of view of the coexistence of the *trans* and the *gauche* molecules the temperature-dependence of the dipole moment would be accounted for by the change of equilibrium ratio of the *trans* and the *gauche* molecules (Watanabe, Mizushima and Morino[31a]), since there is a large difference between the moment values of these two kinds of molecules, the *trans* form having no moment and the *gauche* form having a large moment.

Let $N_t$ and $\mu_t$ be the number and dipole moment of the *trans* molecules and $N_g$ and $\mu_g$ be the corresponding quantities of the *gauche* molecules. Then the mean moment is calculated as

$$m^2 = \frac{\mu_t{}^2 N_t + \mu_g{}^2 N_g}{N_t + N_g} \tag{2.1}$$

In equilibrium we have

$$\frac{N_g}{N_t} = 2 \frac{f_g}{f_t} e^{-\frac{\Delta E}{RT}} = \omega\, e^{-\frac{\Delta E}{RT}} \tag{2.2}$$

where $\Delta E$ is the energy difference between the *trans* and the *gauche* minima, $f_t$ and $f_g$ are the vibrational and rotational partition functions of the *trans* and the *gauche* molecules and the factor 2 is introduced to account for the optical isomerism of the *gauche* form.* From Equations (2.1) and (2.2) we have

$$m^2 = \frac{\mu_t{}^2 + \mu_g{}^2\, \omega\, e^{-\frac{\Delta E}{RT}}}{1 + \omega\, e^{-\frac{\Delta E}{RT}}} \tag{2.3}$$

As the symmetry number is 2 for both of the *trans* and *gauche* forms, we have

$$\omega = 2\frac{f_g}{f_t} = 2\,\frac{(I_1 I_2 I_3)_g{}^{\frac12}\, \Pi\left(1 - e^{-\frac{h\nu_g}{kT}}\right)^{-1}}{(I_1 I_2 I_3)_t{}^{\frac12}\, \Pi\left(1 - e^{-\frac{h\nu_t}{kT}}\right)^{-1}} \tag{2.4}$$

---

* There are two kinds of *gauche* forms, each of them being the mirror image of the other. Both of them have the same normal frequencies and moments of inertia, but they cannot coincide with each other. (See Fig. 2.2.)

31 a. I. Watanabe, S. Mizushima and Y. Morino, Sci. Pap. Inst. Phys. Chem. Res. Tokyo, **39**, 401 (1942).

where $I_1$, $I_2$ and $I_3$ are the principal moments of inertia and $\nu_t$ and $\nu_g$ are the normal frequencies of the *trans* and *gauche* molecules. Putting C–H $= 1.09$ Å, C–C $= 1.54$ Å, and C–Cl $= 1.76$ Å, we have

$$(I_1 I_2 I_3)_t = 2.012 \times 10^6$$
$$(I_1 I_2 I_3)_g = 2.647 \times 10^6 \tag{2.5}$$

in atomic weight-Angstrom units.

As to the normal vibrations, the skeletal frequencies (in $cm^{-1}$) are assigned as follows:*

$$\textit{trans} \text{ molecule} \begin{cases} \delta_s = 300, & \delta_a = 223, & \nu_s\,(C\text{–}Cl) = 754 \\ \nu_a\,(C\text{–}Cl) = 709, & & \nu_s\,(C\text{–}C) = 1052 \end{cases} \tag{2.6}$$

$$\textit{gauche} \text{ molecule} \begin{cases} \delta_1 = 265, & \delta_2 = 411, & \nu_1\,(C\text{–}Cl) = 654 \\ \nu_2\,(C\text{–}Cl) = 677, & & \nu\,(C\text{–}C) = 1031 \end{cases} \tag{2.7}$$

where $\delta$ and $\nu$ are the bending and stretching frequencies and the subscripts $s$ and $a$ refer to the symmetric and antisymmetric vibrations to the center of symmetry. Lacking the data to determine the torsional frequencies about the C–C axis, we assume them to be practically the same for both of the *trans* and *gauche* molecules. Further we consider that the hydrogen vibrations of these two molecular species make, in all, the same contributions to the partition functions.** Then we can evaluate at once the values of $\omega$ at different temperatures as shown in Table 2.5.

We can now determine the energy difference $\Delta E$ between the two molecular species, using the observed values $m$ of mean moment (see Table 2.6). Putting the moment $\mu_t$ of the *trans* molecule as zero, we can calculate the values of $\Delta E$ and $\mu_g$ from Equation (2.3) as follows:

$$\Delta E = 1.21 \text{ kcal./mole.}$$
$$\mu_g = 2.55 \times 10^{-18} \text{ e.s.u.} \tag{2.8}***$$

TABLE 2.5

The value of $\omega = 2 f_g / f_t$ at different temperatures

| $T$ (°K) | $\omega$ |
|---|---|
| 298.2 | $1.90_2$ |
| 400.0 | $1.83_9$ |
| 500.0 | $1.80_6$ |

---

* For the assignment of these skeletal frequencies see Section 6 of Part II.

** The assignment of hydrogen vibrations will be made in Section 9, Part II. As these vibrations do not make significant contribution to the partition function, we calculate their contribution only approximately.

*** The value of $P_E + P_A$ was taken as 23.9 c. c. This was calculated from the dielectric constant of solid where we have only the *trans* molecule with no permanent moment. See also Section 3, Part II.

However, the value of the moment of the *trans* molecule is not exactly zero, since it exerts a torsional oscillation about the C–C axis. This frequency of the *trans* molecule is inactive in the Raman effect, but the corresponding one for the *gauche* molecule is active. We have already stated that these frequencies were assumed to be the same for both molecular species. If, then, we provisionally assign these frequencies to the Raman line observed at 125 cm$^{-1}$* we can evaluate the curvature of the potential curve about the *trans* minimum and obtain at once the mean moments of the *trans* molecule at various temperatures, e. g. at 25° C we have

$$\mu_t = 0.17 \times 10^{-18} \text{ e.s.u.}$$

Putting these values of $\mu_t$ into equation (2.3), we have

$$\left. \begin{array}{l} \varDelta E = 1.22 \text{ kcal./mole} \\ \mu_g = 2.54 \times 10^{-18} \text{ e.s.u.} \end{array} \right\}$$

These values are practically the same as those of Equation (2.8).

TABLE 2.6

Dipole moment of 1,2-dichloroethane in the gaseous state. (Zahn,[31b] Watanabe, Mizushima & Morino[31a])

| $T$ (°K) | $m$ (D) |
|---|---|
| 305.0 | 1.12 |
| 307.3 | 1.18 |
| 337.4 | 1.21 |
| 341.0 | 1.24 |
| 353.2 | 1.25 |
| 376.3 | 1.32 |
| 384.8 | 1.34 |
| 411.6 | 1.38 |
| 419.0 | 1.40 |
| 457.0 | 1.45 |
| 479.8 | 1.48 |
| 484.8 | 1.48 |
| 543.7 | 1.54 |

The values of $\varDelta E$ and $\omega$ being thus determined, we can calculate from Equation (2.3) the equilibrium ratio of these two molecular species at any temperature: e. g.

$$N_g : N_t = 0.34 : 1 \text{ for the vapour at the boiling point.}$$

Similar measurements have been made on 1,2-dibromoethane and the following value of the energy difference has been obtained:

$$\varDelta E = 1.4 \text{ kcal./mol.}$$

In Section 5 we have stated that 1,1,2,2-tetrachloroethane, $Cl_2HC–CHCl_2$, exists in staggered molecular forms, of which the *trans* form has no moment and the *gauche* form has a considerable moment (see Fig. 2.4). Smyth and

---

* As will be explained in Section 9, this frequency corresponds to the torsional frequency of the *gauche* form.

---

31 b. C. T. Zahn, Phys. Rev. **38**, 521 (1931).

McAlpine,[32] Mizushima, Morino and Kozima[33] and Thomas and Gwinn[34] measured the dipole moment of this substance at different temperatures. In contradistinction to the case of 1,2-dichloroethane or 1,2-dibromoethane the dipole moments of 1,1,2,2-tetrachloroethane are almost independent of temperature. The last-mentioned investigators discussed the experimental data in detail with a survey of the C–Cl bond moments for a number of chlorinated compounds and concluded that the energy difference between the two kinds of the staggered configurations in the gaseous state is very small. They also studied the dipole moment of 1,1,2-trichloroethane $Cl_2HC-CH_2Cl$ from which they calculated the energy difference between the rotational isomers as $2.3 \sim 4.0$ kcal./mol. (Fig. 2.6). On the other hand Sheppard and Szasz[24] consider that such a large energy difference is unexpected in view of the small energy difference between the rotational isomers of 2-methylbutane, $(CH_3)_2HC-CH_2 \cdot CH_3$ which has a similar skeletal structure to that of 1,1,2-trichloroethane. However,

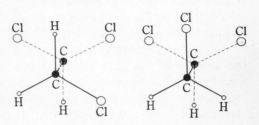

Fig. 2.6. The stable configurations of 1,1,2-trichloroethane.

we must see that although the repulsive potential due to the steric effect would be similar in these two compounds, there will be a large difference in electrostatic potential between them, because the C–Cl bond has a large dipole moment. This may make the energy difference between the rotational isomers of 1,1,2-trichloroethane much larger than that of 2-methylbutane*.

Now that the 1,2-dihalogenoethanes in the liquid state should be regarded as the equilibrium mixtures of the *trans* and the *gauche* isomers, these substances may show some anomalous dielectric behavior as compared with the ordinary substances consisting of uniform molecules. As an example let us explain an interesting observation of Denison and Ramsey[35] on the

---

*Recently Kuratani and Mizushima have found the value of energy difference between the rotational isomers of 1,1,2-trichloroethane in the gaseous state to be 2.9 kcal./mole by the infrared measurement.

32. C. P. Smyth and K. B. McAlpine, J. Am. Chem. Soc., 57, 979 (1935).

33. S. Mizushima, Y. Morino and K. Kojima, Sci. Pap. Inst. Phys. Chem. Res. Tokyo, 29, 111 (1936).

34. J. R. Thomas and W. D. Gwinn, J. Am. Chem. Soc., 71, 2785 (1949).

35. J. T. Denison and J. B. Ramsey, J. Chem. Phys., 18, 770 (1950).

dissociation constant, $K$, of an electrolyte in a solvent of dielectric constant $\varepsilon$ *. They found that there is a large difference in the $K$-values of a given salt when this is dissolved in 1,1-dichloroethane, $Cl_2HC-CH_3$ and in 1,2-dichloroethane, $ClH_2C-CH_2Cl$. Since the dielectric constants of these solvents at ordinary temperature are approximately equal ($\varepsilon = 10$), this large difference seems at first sight quite anomalous. However, they could show that this large difference is due to a difference between the macroscopic dielectric constant of 1,2-dichloroethane and its effective dielectric constant, i. e. the dielectric constant which is effective in determining the stability of the associated ion pair of a salt in that solvent. The existence of the rotational isomers, the polar *gauche* and non-polar *trans* isomers, of 1,2-dichloroethane leads to the conclusion that an ion in this solvent will attract preferentially the polar *gauche* molecules and thereby produce an enhancement of the relative population of this form in the close vicinity of the ion above the average in the pure solvent. This long-range ion-dipole interaction would make the effective dielectric constant in 1,2-dichloroethane greater than the macroscopic dielectric constant of this solvent and thus cause the observed $K$-value to be greater than that expected from the macroscopic dielectric constant. It is evident that in the case of 1,1-dichloroethane the molecule is in one form and there is no difference between the macroscopic and the effective dielectric constants.

## 7. Energy Difference between the Rotational Isomers

In the preceding section we have obtained the energy difference between the *trans* and the *gauche* molecules on the basis of the dielectric constant measurement. However, as such an experimental method is only applicable to the gaseous state or to dilute nonpolar solutions, we are interested in some other techniques by which we can obtain the energy difference between the rotational isomers in the liquid state or in polar solutions. Such methods are provided by the measurements of the Raman and infrared spectra.

Let us first consider the Raman data. The scattered intensity per molecule in the liquid state should not be much different from that in the gaseous state, so that we can determine the change in the equilibrium ratio of rotational isomers from the change of the intensity ratio of two lines, of which one is assigned to the *trans* molecule and the other to the *gauche* molecule. As an example let us consider a pair of Raman lines of 1,2-di-

---

* For the relation between $K$ and $\varepsilon$, see N. Bjerrum, Kgl. Danske Vidensk. Selskab. **7**, No. 9 (1926).

chloroethane at 754 cm.$^{-1}$ and at 654 cm.$^{-1}$, the former being assigned to the *trans* molecule and the latter to the *gauche* molecule. Morino, Watanabe and Mizushima found the intensity ratio I(754)/I(654) for the vapor at 170° C to be 5 which is reduced to 1.8 in the liquid at 25° C.[36] Now that we have evaluated the equilibrium ratio in the gaseous state from the dielectric data we can obtain the ratio in the liquid state from the change of intensity ratio stated above. We have for the liquid at 25° C:

$$N_g : N_t = 1.3 : 1 \qquad (2.9)$$

This means that the energy difference between the rotational isomers of 1,2-dichloroethane is much smaller in the liquid state than that in the gaseous state. Therefore, if we measure the temperature dependence of the intensity ratio of the two lines stated above, we shall not find a remarkable change. Actually Morino, Watanabe and Mizushima found the intensity ratio I(754)/I(654) as 1.8 at 25 °C and as 2.0 at 150 °C.[36] In conformity with this result Rank, Kagarise and Axford[37] obtained an energy difference of zero in the liquid state, using a photoelectric Raman apparatus.

The change of the absorption intensity of the infrared bands has been measured and from this the energy difference between the rotational isomers can also be obtained. Let $I$ and $I_0$ be the intensity of the transmitted and the incident radiations, $\varkappa$ the molecular absorption coefficient, $N$ the number of molecules per c. c., and $l$ the thickness of the absorption layer, we have

$$\ln \left( \frac{I}{I_0} \right) = - \varkappa N l$$

From the intensity measurements for two absorption bands, of which one is assigned to *trans* molecule and the other to the *gauche* molecule, the energy difference is obtained as follows:

$$\frac{\ln (I_t/I_{ot})}{\ln (I_g/I_{og})} = \frac{\varkappa_t N_t}{\varkappa_g N_g} = \frac{\varkappa_t}{\varkappa_g} \frac{f_t}{2 f_g} e^{\frac{\varDelta E}{R T}} \qquad (2.10)$$

Shimanouchi, Tsuruta and Mizushima[18] observed the molar absorption coefficient for the two absorption bands, one at 1232 cm.$^{-1}$ assigned to the *trans* molecule of 1,2-dichloroethane and the other at 1291 cm.$^{-1}$ assigned to the *gauche* molecule and found

$$\varDelta E = 1.03 \, \text{kcal./mol},$$

36. Y. Morino, I. Watanabe and S. Mizushima, Sci. Pap. Inst. Phys. Chem. Res. Tokyo, **39**, 396 (1942).

37. D. H. Rank, R. E. Kagarise and D. W. E. Axford, J. Chem. Phys., **17**, 1354 (1949).

which agrees with that obtained from the dielectric constant measurement within the limit of experimental error. A few years later Bernstein[38] made a similar measurement independently and obtained results in good agreement with those given above:

$$\varDelta E = 1.32 \text{ kcal./mol.}$$

Shimanouchi, Tsuruta and Mizushima[18] applied the same method to the intensity measurement of the two bands of the gaseous 1,2-dibromoethane at 1187 cm.$^{-1}$ (*trans*) and at 1251 cm.$^{-1}$ (*gauche*) and found the energy difference to be

$$\varDelta E = 1.45 \text{ kcal./mole}$$

which is also in good agreement with the value obtained by Watanabe, Mizushima and Morino from the dielectric constant measurement:

$$\varDelta E = 1.40 \text{ kcal./mole}$$

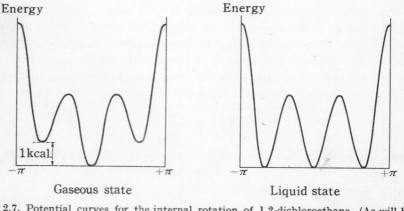

Fig. 2.7. Potential curves for the internal rotation of 1,2-dichloroethane. (As will be shown in Section 10 the height of the lower barriers is not much different from 3 kcal./mole. The other barrier is much higher than these.)

For the liquid 1,2-dibromoethane Rank, Kagarise and Axford[37] found

$$\varDelta E = 0.74 \text{ kcal./mole}$$

from the Raman measurement. This is in good agreement with the value

$$\varDelta E = 0.76 \text{ kcal./mole}$$

38. H. J. Bernstein, J. Chem. Phys., 17, 258 (1949).

obtained by Morino, Mizushima, Kuratani and Katayama[39] who observed the temperature dependence of the intensity of two Raman lines at 660 cm.$^{-1}$ (*trans*) and at 551 cm.$^{-1}$ (*gauche*). The same authors made an intensity

TABLE 2.7

The energy difference between the rotational isomers of 1,2-dihalogenoethanes in kcal./mol.

| Substance | State | Energy difference | Reference |
|---|---|---|---|
| $ClH_2C–CH_2Cl$ | Gas | 1.2  (dielectric const.)<br>1.0  (infrared)<br>1.1  (infrared)<br>1.3  (heat capacity) | Watanabe, Mizushima and Morino.[31a]<br>Shimanouchi, Tsuruta and Mizushima[18]<br>Bernstein[38]<br>Gwinn and Pitzer[40] |
| | Liquid | 0  (Raman)<br>0  (Raman)<br>0  (Raman)<br>0  (infrared)<br>0  (dielectric const.) | Morino, Watanabe and Mizushima[36]<br>Rank, Kagarise and Axford[37]<br>Gerding and Meerman[43]<br>Kuratani and Mizushima[44]<br>Watanabe, Mizushima and Masiko[45] |
| $BrH_2C–CH_2Br$ | Gas | 1.45 (infrared)<br>1.77 (infrared)<br>1.4  (dielectric const.) | Shimanouchi, Tsuruta and Mizushima[18]<br>Bernstein[41]<br>Watanabe, Mizushima and Morino.[31a] |
| | Liquid | 0.74 (Raman)<br>0.76 (Raman)<br><br>1.39 (Raman)<br>2.42 (Raman)<br>0.65 (infrared)<br><br>0.73 (infrared) | Rank, Kagarise and Axford[37]<br>Morino, Mizushima, Kuratani and Katayama[39]<br>Gerding and Meerman[43]<br>Aronov, Tatevski and Frost[42]<br>Morino, Mizushima, Kuratani and Katayama[39]<br>Kuratani and Mizushima[44] |

39. Y. Morino, S. Mizushima, K. Kuratani and M. Katayama, J. Chem. Phys., **18**, 754 (1950).

40. W. D. Gwinn and K. S. Pitzer, J. Chem. Phys., **16**, 303 (1948).

41. H. J. Bernstein, J. Chem. Phys., **18**, 897 (1950).

42. O. L. Arnov, V. M. Tatevski and A. V. Frost, Doklady Acad. Nauk, U.S.S.R., **60**, 387 (1948).

43. H. Gerding and P. G. Meerman, Rec. Trav. Chim., **61**, 523 (1942).

44. K. Kuratani and S. Mizushima, unpublished.

measurement of the infrared absorption bands at 1188 cm.$^{-1}$ (*trans*) and at 1246 cm.$^{-1}$ (*gauche*) at different temperatures and obtained

$$\varDelta E = 0.65 \text{ kcal./mole}$$

In Table 2.7 are summarized the data for the energy difference between the rotational isomers of 1,2-dichloroethane and 1,2-dibromoethane in the gaseous and the liquid states.* In Fig. 2.7 are shown the general form of the potential curve for the internal rotation of 1,2-dihalogenoethane obtained on the basis of the experimental data so far described.

As stated in the preceding sections the change in these potential curves must be due to the interaction between molecules. Now what is the force playing the most important role in the interaction stated above? Since the decrease of the energy difference in the liquid state must arise from the different behavior between the *trans* and the *gauche* molecules (the dipole moment of the *trans* molecule being zero and that of the *gauche* molecule large), it is very probable that the most important force is the electrostatic force. Watanabe, Mizushima and Masiko calculated the energy change of the free molecule in the liquid state from such an electrostatic consideration.[45]

According to Onsager, when a dipole of moment $\mu$ is transferred from a vacuum ($\varepsilon = 1$) into a medium of dielectric constant $\varepsilon$, it loses potential energy by an amount of

$$E_r = \frac{(\varepsilon - 1)}{2\,\varepsilon + 1} \frac{\mu^2}{a^3} \tag{2.11}$$

---

* The calculation of the energy difference in the liquid state from the values of dielectric constant has been made through the Onsager equation:

$$m^2 = \frac{9\,k\,T}{4\,\pi\,N} \frac{(\varepsilon - n^2)\,(2\,\varepsilon + n^2)}{\varepsilon\,(n^2 + 2)^2} \frac{M}{d}$$

where $\varepsilon$ is the dielectric constant of the liquid. Putting

$$\frac{n^2 - 1}{n^2 + 2} \frac{M}{d} = P_E + P_A = 23.9 \text{ c. c.} \quad (1,2\text{-dichloroethane})$$

and using the values of dielectric constant measured by Watanabe, Mizushima and Masiko,[45] we obtain the values of $m^2$ or the square of the apparent moment. From these values we can calculate the energy difference just as we did in the case of dielectric constant in the gaseous state. There may be some objections with regard to the application of the Onsager equation to such a problem, but it is interesting that we have found a result compatible with those obtained by the more reasonable methods explained in this section.

---

45. I. Watanabe, S. Mizushima and Y. Masiko, Sci. Pap. Inst. Phys. Chem. Res. Tokyo, **40**, 425 (1943). See also, S. Mizushima and H. Okazaki, J. Am. Chem. Soc., **71**, 3411 (1949).

where $a$ is the molecular radius. Therefore, the *trans* isomer, having no dipole, is subjected to no energy change, while the *gauche* isomer of moment $\mu$ loses energy by an amount given by Eq. (2.11). In consequence the *gauche* form becomes more stable in the liquid state than in the gaseous state. Putting the values of dipole moment ($\mu = 2.55$ D) and dielectric constant ($\varepsilon = 9.87$) of 1,2-dichloroethane into the right-hand side of Eq. (2.11) we have

$$E_r = 1.0 \text{ kcal./mole.}$$

This is of just the magnitude which we have found as the difference of $\Delta E$ between the gaseous and liquid states shown in Table 2.7 and, therefore, it is very probable that the electrostatic force plays the most important role in the change of the energy difference between the rotational isomers stated above.

Similar calculations have been made for 1,2-dibromoethane with the result

$$E_r = 0.5 \text{ kcal./mole}$$

which is also compatible with the experimental results shown in Table 2.7.[39]

In Section 2 we have seen that the energy parameter $V_0$ shown in Table 1.2 depends considerably upon the nature of the solvent in which the dipole measurement is made and $V_0$ has been shown to decrease in a solvent of higher dielectric constant. In terms of rotational isomer theory this means that the energy difference between the *trans* and the *gauche* molecules becomes less, or that the relative number of *gauche* molecules increases in a medium of higher dielectric constant, which is quite compatible with the electrostatic view considered above.

In this connection the intensity measurement of Raman lines of 1,2-di-chloroethane and 1,2-dibromoethane made by Mizushima, Morino and Higasi are of interest.[46] They observed the change of intensity ratio $I(\nu_1)/I(\nu_2)$ of two lines with solvent, where $\nu_1$ is assigned to the *trans* molecule and $\nu_2$ to the *gauche* molecule and found that $I(\nu_1)/I(\nu_2)$ increased from 1.2 to 5 in 1,2-dichloroethane and from 7 to 15 in 1,2-dibromoethane, when the dielectric constant of the solvent was changed from 33 ($CH_3OH$) to 2 ($C_6H_{14}$).

This experimental result also supports the view that the interaction between molecules in the liquid state is mainly electrostatic in nature in

---

46. S. Mizushima, Y. Morino and K. Higasi, Physik, Z., **35**, 905 (1934); Sci. Pap. Inst. Phys. Chem. Res. Tokyo, **25**, 159 (1934).

TABLE 2.8

$I\ (\nu_1)/I\ (\nu_2)$  in different solvents[46]

| Solvent | | $ClH_2C–CH_2Cl$ $\nu_1 = 754$ cm.$^{-1}$ $\nu_2 = 654$ cm.$^{-1}$ | $BrH_2C–CH_2Br$ $\nu_1 = 660$ cm.$^{-1}$ $\nu_2 = 551$ cm.$^{-1}$ |
|---|---|---|---|
| $CH_3OH$ | $\varepsilon = 33$ | 1.2 | 7 |
| $C_2H_5OH$ | $\varepsilon = 25$ | 1.3 | 8 |
| $C_2H_4Cl_2$ | $\varepsilon = 10$ | 1.4 | — |
| $C_2H_4Br_2$ | $\varepsilon = 4.8$ | — | 11 |
| $(C_2H_5)_2O$ | $\varepsilon = 4.4$ | 2.1 | 11 |
| $C_6H_{14}$ | $\varepsilon = 2.0$ | 5 | 15 |

1,2-dihalogenoethanes.* Recently Kuratani measured the energy difference between the *trans* and the *gauche* forms in different solvents by infrared methods with essentially the same result as above.[47]

Wada improved the electrostatic model stated above by considering a spheroidal cavity with dipoles located at its foci and using the moment value observed in solutions. In this way he could explain quantitatively the change of the energy differenc between the *trans* and the *gauche* molecules with the dielectric constant of the solvent.[48]

In the preceding section we have stated that the energy difference between the rotational isomers of 1,1,2,2-tetrachloroethane in the gaseous state is very small. This does not agree with the result of the Raman intensity measurement of Langseth and Bernstein[49] and Kagarise and Rank[26] who found the energy difference in the liquid state to be 1.1 and 0.9 kcal./mole., respectively. A value of energy difference not much different from these was also obtained by Naito et al.[27] This apparent inconsistency can be explained, if we can assign one of the rotational isomers of 1,1,2,2-tetrachloroethane to the *trans* form with no moment and the other to the *gauche* form with large moment,[50] because the extra stabilization of the *gauche* form in

---

* In this case benzene was found to give an abnormal result just as in the case of the dipole measurement. The value of $I\ (\nu_1)/I\ (\nu_2)$ was found to be much smaller than would be expected from the dielectric constant of benzene; in other words the relative number of the *gauche* molecules is unexpectedly large in benzene solution.

47. K. Kuratani, Rep. Inst. Sci. Tech. Tokyo Univ., 6, 221 (1952).
48. A. Wada, private communication.
49. A. Langseth and H. J. Bernstein, J. Chem. Phys., 8, 410 (1940).
50. N. Sheppard and G. J. Szasz, J. Chem. Phys., 18, 145 (1950).

the liquid state will result in a considerable change in energy difference between the *trans* and the *gauche* molecules, both of which have almost equal stability in the gas phase. (Langseth and Bernstein concluded the stable forms to be the eclipsed *cis* form and the *gauche* form. However, in order to reconcile their value of energy difference with the explanation stated above, the stable forms should be assigned, as stated in Section 5, to the *trans* and the *gauche* forms, where the *gauche* form becomes more stable in the liquid state.)

## 8. Electron Diffraction

Electron diffraction investigations provide us with further experimental evidence to support the view of the rotational isomerism in 1,2-dihalogeno-ethanes.

According to the experimental results of the Raman effect, infra-red absorption and dielectric constant, the molecules of hexachloroethane, pentachloroethane, asymmetrical tetrachloroethane, etc. exist in one form which is the staggered form, in which the projection of one C–Cl or C–H line upon the plane perpendicular to the C–C axis is midway between those of the two C–Cl lines in the other half of the molecule. In Fig. 2.8 is shown the theoretical intensity curve of hexachloroethane, $Cl_3C–CCl_3$, obtained from the usual expression for the intensity of scattered electrons:

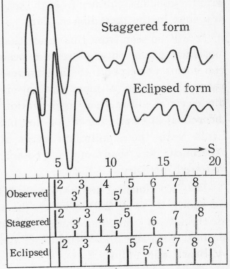

Fig. 2.8. Theoretical intensity curves of hexachloroethane and positions of observed maxima.

$$I = \sum_i \sum_j Z_i Z_j \frac{\sin s\, r_{ij}}{s\, r_{ij}}$$

where $r_{ij}$ is the distance from the $i$-th to the $j$-th atom and $Z$ is atomic number. $s$ is related to the wave length $\lambda$ of the de Broglie wave and the scattering angle $\theta$ through the following formula:

$$s = \frac{4\pi}{\lambda} \sin \frac{\theta}{2}$$

In the same figure are shown the positions of maxima obtained from this intensity curve, which are in good agreement with those observed in the electron diffraction experiment of Morino, Yamaguchi and Mizushima.[51] The theoretical intensity curve calculated for the eclipsed form, in which the three C–Cl bonds eclipse the three C–Cl bonds in the other half of the molecule as viewed along the central C–C axis, is not at all compatible with the experimental result.

For pentachloroethane, $Cl_2HC–CCl_3$ and asymmetrical tetrachloroethane, $ClH_2C–CCl_3$ the theoretical intensity curves calculated for the staggered form have also been shown to be in agreement with the maxima and minima observed in the electron diffraction experiments of the same authors.[51] All the experimental data on electron diffraction so far obtained for halogeno-ethanes with only one molecular configuration are shown to be compatible with the conclusion drawn from the spectroscopic and dielectric investigations described in the preceding sections.

It will be seen from this result that in 1,2-dihalogenoethanes the stable molecular configurations are the *trans* and the *gauche* forms, since these two forms correspond to the staggered forms of the molecules described above. However, a more straightforward proof for this conclusion from the electron diffraction investigation can be made by calculating the theoretical intensity curve for the mixture of the *trans* and the *gauche* forms with the equilibrium ratio at the temperature at which the experiment was made and by comparing this curve with the measured scattered intensity.

The experimental results obtained for the vapor of 1,2-dichloroethane are shown in Table 2.9, in which the second column refers to the s-values for the intensity maxima or the intensity minima observed by Yamaguchi, Morino, Watanabe and Mizushima[52] and the third column to those observed by Beach and Palmer.[53]

The theoretical intensity curves are calculated taking the atomic distances as $r_{C–C} = 1.54$ Å, $r_{C–Cl} = 1.76$ Å and $r_{C–H} = 1.09$ Å and assuming all the valence angles tetrahedral. In Fig. 2.9 are shown the curves calculated for the *trans*, *gauche*, and *cis* forms, of which the *trans* curve corresponds to

51.  Y. Morino, S. Yamaguchi and S. Mizushima, Sci. Pap. Inst. Phys. Chem. Res. Tokyo, **42**, 5 (1944).

52.  S. Yamaguchi, Y. Morino, I. Watanabe and S. Mizushima, Sci. Pap. Inst. Phys. Chem. Res. Tokyo, **40**, 417 (1943).

53.  J. Y. Beach and K. J. Palmer, J. Chem. Phys. **6**, 639 (1938).

the experimental result most closely.  Hence we have to consider that the *trans* form is predominant in the vapor of 1,2-dichloroethane at 14° C, at which temperature the experiments in our laboratory were made.

The third maximum at $s = 6.003$, the existence of which was first shown in the experiments of our laboratory, exists in the theoretical curves for the pure *trans* isomer and for the mixtures of the *trans* and the *gauche* isomers, but is absent in the curves for the mixtures of the *trans* and *cis* isomers, in which the *cis* isomer exists in reasonable amount. Hence, so far as this maximum is concerned, all the theoretical curves for the mixtures of the *trans* and *gauche* isomers are compatible with the experi-

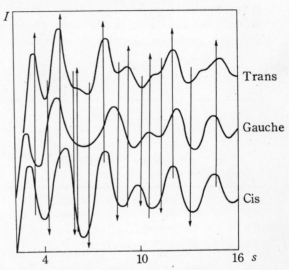

Fig. 2.9. Theoretical intensity curves of 1,2-dichloro-ethane in the *trans*, *gauche* and *cis* forms and the observed intensity maxima (↑) and minima (↓).

mental result, but there are no grounds for admitting the existence of a mixture of the *trans* and *cis* isomers.

The fifth maximum at $s = 9.205$ is only present in the isomeric mixture in which composition of the *gauche* form ranges from 0 to 30%, hence this isomer should exist in the isomeric mixture in an amount less than 30%. Among the several intensity curves calculated for the mixtures in this range of composition, the 19% curve corresponds most closely to the experiment. This is just the composition which we would expect at 14° C from the energy difference between the rotational isomers of 1,2-dichloroethane explained in the preceding section.  It would take up too much space to go into details about the discussion of the other intensity maxima, but the explanation given above should be sufficient to show that the experimental results of the electron diffraction investigation are also compatible with the conclusions from the optical and dielectric studies.

Quite recently Ainsworth and Karle have made more detailed diffraction investigation on 1,2-dichloroethane with essentially the same result as stated above.[54] They found the azimuthal angle of the *gauche* form to be 109° ± 5° from the *trans* position. This would mean that the repulsion between two chlorine atoms is greater than that between chlorine and hydrogen atoms, attached to different movable groups, so that the distance between two chlorine atoms becomes greater and that between chlorine and hydrogen atoms smaller than those at the azimuthal angle 120°.

TABLE 2.9

*s*-values for the observed maxima and minima ₹

| | Maxima | | | Minima | |
|---|---|---|---|---|---|
| No. | Yamaguchi, Morino, Watanabe & Mizushima | Beach and Palmer | No. | Yamaguchi, Morino, Watanabe & Mizushima | Beach and Palmer |
| 1 | 3.205 (strong) | 3.160 | 1 | 4.141 | 4.136 |
| 2 | 4.940 (strong) | 4.936 | 2 | 5.902 | — |
| 3 | 6.003 (weak) | — | 3 | 6.664 | 6.668 |
| 4 | 7.699 (strong) | 7.679 | 4 | 8.591 | 8.594 |
| 5 | 9.205 (medium) | 9.194 | 5 | 9.993 | 9.964 |
| 6 | 10.598 (weak) | 10.591 | 6 | 11.305 | 11.302 |
| 7 | 12.051 (strong) | 12.041 | 7 | 13.189 | 13.183 |
| 8 | 14.813 (medium) | 14.815 | | | |

The X-ray diffraction investigation on crystalline 1,2-dichloroethane was made by Milberg and Lipscomb[55] who found exactly the same result as that of the spectroscopic investigations of Mizushima, Morino et al.[2, 3] viz. that all the molecules in the crystalline state are in the *trans* form.* The heat capacity curve of this substance by Pitzer[56] shows a large peak indicating a transition near 177° K in the solid state. However, the Raman measurement of Mizushima and Morino shows definitely that the molecular form (the *trans* form) does not change at all on crossing the transition point. Accordingly, if we consider the excitation of a molecular motion above the transition point, it should be such a kind of motion which does not change the molecular configuration appreciably; for example, an overall rotation about, (approxi-

---

* The X-ray study of crystalline 1,2-diiodoethane was made by Klug who also found only the *trans* form in the solid state. See H. P. Klug, J. Chem. Phys., **3**, 747 (1935).

---

54. J. Ainthworth and J. Karle, J. Chem. Phys., **20**, 425 (1952).
55. M. E. Milberg and W. N. Lipscomb, Acta Cryst., **4**, 369 (1951).
56. K. S. Pitzer, J. Am. Chem. Soc., **62**, 331 (1940).

mately) the Cl...Cl axis. This is just what was suggested by a nuclear magnetic resonance study made by Gutowsky and Pake.[57]

Mizushima and Morino[3] observed two low frequency Raman lines at 53 and 74 cm.$^{-1}$ below the transition point and only one line at 66 cm.$^{-1}$ above that point. This is the only change of the Raman spectra of crystalline 1,2-dichloroethane which takes place on crossing the transition point and can be explained in conjunction with what has been stated above (Ichishima and Mizushima[58]). The two lines observed below the transition point can be assigned to the rotatory vibration about the two axes, both perpendicular to the molecular axis. Since the moments of inertia about these two axes are almost equal to each other, the appearance of these two Raman lines would be due to the difference of the restoring forces for these two motions. Since above the transition point the free rotation about the molecular axis sets in, the restoring forces will become equal to each other and, accordingly, we expect only one Raman line assigned to the rotatory vibration about the axis perpendicular to the molecular axis. Mizushima and Morino also observed the change of low frequency Raman lines of crystalline 1,2-dibromoethane on crossing the transition point, which can be explained in the same manner as in the case of 1,2-dichloroethane.

## Summary of Chapter II

The number of the Raman lines of the 1,2-dihalogenoethanes, $XH_2C-CH_2X$ in the liquid and the gaseous states is too large to be accounted for as arising from a single molecular form. In the solid state, however, these Raman lines are materially reduced in number and the spectra can be explained by assuming a single molecular form. From this observation it is possible to pick out separate spectra for the more stable form persisting in the solid state and the less stable form which can exist only in the liquid and gaseous states. The number of Raman lines assigned to the more stable form has been compared with the number of Raman lines which would be expected from the selection rules for conceivable configurations of the 1,2-dihalogenoethane molecule. This comparison shows clearly that the more stable form is the *trans* form which is a staggered configuration with a center of symmetry and in which the two halogen atoms are as far apart as possible. Therefore, all the Raman lines observed in the solid state should be assigned to the symmetric vibrations of the *trans* form and their frequencies should not

57. H. S. Gutowsky and G. E. Pake, J. Chem. Phys., 18, 162 (1950).
58. I. Ichishima and S. Mizushima, J. Chem. Phys., 18, 1420 (1950).

agree with the infrared absorption frequencies which are assigned to the antisymmetric vibrations. This was shown to be actually the case.

The less stable form which is found in the liquid and gaseous states with the *trans* form is the other staggered variety obtained by rotating one end of the *trans* form through an angle of 120°. This was shown by the application of the product rule for rotational isomers and by the Raman measurement made on deuterium-substituted dihalogenoethanes. The fact that $ClH_2C-CCl_3$ and $Cl_2HC-CCl_3$ exist in only one molecular form is in good agreement with the above view, since the three staggered forms of these molecules are identical with one another.

Of the two molecular forms referred to above, the *trans* form with center of symmetry has no dipole moment, while the *gauche* form has a large moment. The observed change of apparent moment with temperature can be explained from the change of equilibrium ratio of the *trans* and the *gauche* molecules with temperature. The change of this ratio is evidently related to the energy difference between these two forms and, accordingly, from the temperature dependence of the apparent moment we can evaluate the energy difference. This was found to be 1.2 kcal./mole in the case of gaseous 1,2-dichloroethane. Practically the same value was obtained from the temperature dependence of the intensity ratio of two infrared absorption bands, one assigned to the *trans* form and the other to the *gauche* form.

Similar spectroscopic measurements made in the liquid state show that the equilibrium ratio in this state is different from that in the gaseous state. This is explained as being due to the stabilization of the polar form (the *gauche* form) in the liquid state through an electrostatic interaction.

The electron diffraction pattern of the gaseous 1,2-dichloroethane can also be explained quite reasonably as arising from the mixture of the *trans* and the *gauche* forms with the equilibrium ratio determined from the dielectric and spectroscopic measurements.

Similar spectroscopic and electric measurements have been made for other ethane derivatives and it has been shown that only the staggered forms are stable. There is no possibility for the existence of the stable eclipsed forms which were proposed by some investigators.

The spectroscopic data discussed in this chapter show definitely the existence of rotational isomers and accordingly we have to conclude that in ethane derivatives there is more than one potential minimum in one rotation about the C–C bond as axis. This view is preferable to that stated in the preceding chapter, in which we assumed only one broad potential minimum in one internal rotation.

# CHAPTER III

# Ethane and Its Derivatives II
## (Thermal Measurements and General Discussion)

### 9. Entropy of 1,2-Dihalogenoethanes

We have now a comprehensive knowledge of the configurations and the vibrational frequencies of the molecules of some ethane derivatives from which we can calculate the thermodynamic functions of these substances in the gaseous state.

To a high degree of approximation we can separate these into three parts, corresponding to the separation of the energy into translational, rotational, and vibrational parts. In the case of entropy the contributions of the translational, rotational and vibrational motions, $S_t$, $S_r$ and $S_v$ are expressed as follows:

$$S_t = \frac{3}{2} R \ln M + \frac{5}{2} R \ln T - R \ln P - 2.314 \tag{3.1}$$

$$S_r = \frac{R}{2} \ln (I_1 I_2 I_3 \times 10^{117}) + \frac{3}{2} R \ln T - R \ln \sigma - 0.058 \tag{3.2}$$

$$S_v = R \sum_i \left[ -\ln \left( 1 - e^{-\frac{h \nu_i}{kT}} \right) + \frac{\frac{h \nu_i}{kT}}{e^{\frac{h \nu_i}{kT}} - 1} \right] \tag{3.3}$$

where $R$ denotes the gas constant, $k$ the Boltzmann constant, $h$ the Planck constant, $M$ the molecular weight, and $I_1 I_2 I_3$ the product of principal moments of inertia and $\nu_i$ the normal frequencies, $\sigma$, which is 2 for both the *trans* and *gauche* molecules, is the symmetry number, or the number of indistinguishable positions into which the molecule can be turned by simple rigid rotations.

The products of the principal moments of inertia and the normal frequencies have been already given in Section 6, except for the torsional frequencies about the C–C single bond. Therefore the molecular entropy, without the contribution of the torsional motions, can be evaluated at once through Equations (3.1), (3.2) and (3.3). Let this part of the molecular

51

entropy of the *trans* and the *gauche* molecules be $S_t'$ and $S_g'$, respectively and further let the contribution of the torsional motion be $S''$, assuming the same value for the *trans* and the *gauche* molecules for the sake of simplicity.  Then the entropy of the equilibrium mixture of the *trans* and the *gauche* isomers of 1,2-dihalogenoethane can be expressed as:

$$S = (S_t' + S'') \, x + (S_g' + S'') \, (1 - x) - R \, x \ln x - R \, (1 - x) \ln (1 - x) \quad (3.4)$$

where $x$ is the mole fraction of the *trans* isomer which can be determined from the energy difference between the rotational isomers as described in Section 6.  The calculation of the molecular entropy has been made by Morino, Watanabe and Mizushima for 1,2-dichloroethane[1] and by Odan, Mizushima and Morino for 1,2-dibromoethane,[2] using the molecular data obtained from the spectroscopic and dielectric measurements described in the preceding sections.

The entropy of 1,2-dihalogenoethane can also be obtained on the basis of the third law, if we know all the necessary thermal data down to sufficiently low temperatures.  In the case of 1,2-dichloroethane and 1,2-dibromoethane we have such data from the work of Pitzer on the condensed phase.[3]  Combining the latent heat value obtained in our laboratory with these data we can calculate the third law entropy in the gaseous state.  This we put into $S$ on the left-hand side of Eq. (3.4) in order to obtain the value of $S''$ on the right-hand side of the same equation.  We have

$$S'' = 3.88 \text{ e. u. for the vapor of 1,2-dichloroethane at } 298^\circ \text{ K}$$

and

$$S'' = 3.05 \text{ e. u. for the vapor of 1,2-dibromoethane at } 200^\circ \text{ K}$$

under their own vapor pressures.  From these entropy values we can calculate the frequencies of the torsional motion as 97 cm.$^{-1}$ and 83 cm.$^{-1}$, respectively, assuming the motion to be harmonic.[1,2]

It would now be very interesting to see whether we can observe the frequencies of these torsional motions optically.  In the case of the *trans* molecule, this frequency is antisymmetric to the center of symmetry and, therefore, cannot be observed in the Raman effect.  For the *gauche* molecule, however, this frequency is Raman-active.

---

1. Y. Morino, I. Watanabe and S. Mizushima, Sci. Pap. Inst. Phys. Chem. Res. Tokyo, **40**, 100 (1942).

2. M. Odan, S. Mizushima and Y. Morino, Sci. Pap. Inst. Phys. Chem. Res. Tokyo, **42**, 27 (1944).

3. K. S. Pitzer, J. Am. Chem. Soc., **62**, 331 (1940).

As a line of the frequency of this order of magnitude, we observed for each of the two substances a low frequency line (125 cm.$^{-1}$ for 1,2-dichloroethane and 91 cm.$^{-1}$ for 1,2-dibromoethane) which is assigned to the *gauche* molecule.[1,2] Now that the frequency of the torsional motion has been obtained by the difference method, it is in error by the sum of the error in the molecular calculation and that in the thermal measurement. In addition we have assumed simply that the frequencies for both of the molecular forms are the same. In view of this situation it is not unreasonabe to identify the Raman lines of 125 cm.$^{-1}$ and 91 cm.$^{-1}$ with the torsional frequencies of the *gauche* molecules.

The discussion of entropy given above has taken into account only the *trans* and the *gauche* configurations. This corresponds to a potential curve with fairly sharp minima in the *trans* and the *gauche* positions. Such a curve would be quite reasonable in view of the sharpness of the Raman lines assigned to the *trans* and *gauche* molecules.

In this connection the calculation of entropy made by Gwinn and Pitzer[4] on the basis of a potential of the type,

$$V(\theta) = V_1 \frac{1-\cos\theta}{2} + V_3 \frac{1-\cos 3\theta}{2},$$

is very interesting. After a detailed discussion they selected a potential curve consisting of the two parts:

$$V = 1830 \frac{1-\cos\theta}{2} + 2300 \frac{1-\cos 3\theta}{2}\left(\frac{\text{cal.}}{\text{mole}}\right) \quad 0 < \theta \leqslant 130°$$

and

$$V = \infty$$

by means of which they explained not only the entropy data but also the heat capacity and dipole moment data for 1,2-dichloroethane in the gaseous state.

## 10. The Potential Barrier to Internal Rotation

The most important contribution of the measurements of entropy and specific heat to the study of internal rotation has been made in the case of ethane and other hydrocarbons, because in this case the dielectric measurement is not applicable and, moreover, as explained below, the height of potential barrier hindering internal rotation can be estimated owing to the threefold symmetry of the potential function.

---

4. W. G. Gwinn and K. S. Pitzer, J. Chem. Phys., **16**, 303 (1948).

The contribution to entropy of the torsional motion about the C–C single bond as axis can be obtained in the case of ethane just as in the case of the 1,2-dihalogenoethanes as explained in Section 9. The third law entropy of ethane is determined by the thermal measurements and the molecular entropy is calculated from the spectroscopic data, lacking that of torsional motion. From the difference between these two values of entropy we can obtain the contribution of the torsional motion and, further, the barrier height, assuming a simple form of the potential function:

$$V = V_0 (1 - \cos 3\,\theta) \tag{3.5}$$

where $2\,V_0$ is the height of the peaks above the valleys.

Instead of obtaining the entropy value of ethane itself we may choose an indirect method in which we calculate the equilibrium constant for the hydrogenation of ethylene,

$$C_2H_4 + H_2 = C_2H_6,$$

using the heat of hydrogenation and the spectroscopic and molecular data. Such calculations were made in 1935 by Teller and Topley[5] and independently by Smith and Vaugan[6] who found a discrepancy which would disappear if the height of the barrier were actually 3 kcal./mole, instead of 0.3 kcal./mole previously calculated by Eyring (see Section 1). In the meantime measurements of the third law entropy of tetramethylmethane by Aston and Messerly[7] and of ethane by Witt and Kemp[8] had been made and these results led Kemp and Pitzer[9] to suggest that a barrier of about 3 kcal./mole hindering internal rotation was present in all cases. Following this the important papers of Pitzer[10] and Crawford[11] appeared in which the more rigorous theoretical treatment of the thermodynamic functions for a hindered rotator was given.

Such being the case, it became desirable to remeasure the specific heat of ethane which had earlier been measured by Eucken and Weigert[12] who had found good agreement with the low value of the potential barrier calculated by Eyring.[13]

5.  E. Teller and B. Topley, J. Chem. Soc., **1935**, 876.
6.  H. A. Smith and W. E. Vaugan, J. Chem. Phys., **3**, 341 (1935).
7.  J. C. Aston and G. M. Messerly, J. Chem. Phys., **4**, 391 (1936).
8.  R. K. Witt and J. D. Kemp, J. Am. Chem. Soc., **59**, 273 (1937).
9.  J. D. Kemp and K. S. Pitzer, J. Chem. Phys., **4**, 749 (1936).
10.  K. S. Pitzer, J. Chem. Phys., **5**, 469, 473 (1937).
11.  B. L. Crawford, J. Chem. Phys., **8**, 273 (1940).
12.  A. Eucken and K. Weigert, Z. phys. Chem., **B 23**, 265 (1933).
13.  H. Eyring, J. Am. Chem. Soc., **54**, 3191 (1932).

As stated in the preceding section, we can separate the specific heat into three parts, of which the sum of the translational and the rotational parts is equal to $3R$ except at very low temperatures. The vibrational part can be expressed by the Planck-Einstein formula as:

$$C_v = \sum_{i=1}^{n} \varphi \left( \frac{h v_i}{k T} \right)$$

$$\varphi \left( \frac{h v_i}{k T} \right) = R \frac{\left( \frac{h v_i}{k T} \right)^2 e^{\frac{h v}{k T}}}{\left( e^{\frac{h v_i}{k T}} - 1 \right)^2} \qquad (3.6)$$

where $v_i$ is a normal frequency and $n$ the number of the vibrational modes. If, therefore, we know all the normal frequencies except the torsional frequency, this can be obtained from the experimental value of the specific heat of the gas. We can also determine the height of the potential from this torsional frequency by assuming again the simple form of the potential function as given in Eq. (3.5). Kistiakowsky, Lacher and Stitt[14] made an exact measurement of the specific heat of ethane from which they concluded a barrier height of 2.75 kcal./mole in good agreement with that obtained from the entropy value.

We have good reason to believe that the actual single bond itself does not resist internal rotation appreciably and therefore, the observed phenomena must be associated with the electron pairs attaching other groups, i. e. the C–H bond electrons in ethane. In this connection it would be interesting to see if we can find a hindering potential of zero in such a molecule as dimethyl acetylene $H_3C-C\equiv C-CH_3$ in which the two methyl groups are separated from each other by a comparatively large distance. Osborn, Garner and Yost[15] found agreement between the third law entropy and that calculated from the molecular data on the assumption of free internal rotation. Crawford and Rice[16] and Kistiakowsky and Rice[17] also found a hindering potential of zero for this substance from gaseous heat capacity measurements. Zinc dimethyl and mercury dimethyl with two

14. G. B. Kistiakowsky, J. R. Lacher and F. Stitt, J. Chem. Phys., **6**, 407 (1938); **7**, 289 (1939).

15. D. W. Osborne, C. S. Garner and D. M. Yost, J. Chem. Phys., **8**, 131 (1940).

16. B. L. Crawford and W. W. Rice, J. Chem. Phys., **7**, 437 (1939).

17. G. B. Kistiakowsky and W. W. Rice, J. Chem. Phys., **8**, 618 (1940).

methyl groups far apart in a linear $H_3C$–metal–$CH_3$ skeleton have also been shown to exert free internal rotation by the analysis of the perpendicular bands near $3\mu$ (Boyd, Williams & Thompson).[18]

The question of the hindering potential stated above is one of the most challenging problems in the field of molecular structure and there have been a number of workers who have tried to explain the barrier in terms of valence theory. Among other explanations, that of Lassettre and Dean[19] appears to be the most interesting. They attempt to account for the barrier by an electrostatic model recognizing quadrupole as well as dipole moments for the bonds. In ethane and in other hydrocarbons, the dipole moments are small but, if the bonding electrons are largely concentrated between the nuclei, the quadrupole moments could be large enough to account for the observed barriers. However, if we extend the problem to ethane derivatives or to other more complex molecules, the true explanation of the barrier becomes much complex and there is still no convincing and well established theory which accounts quantitatively for the potential barriers to internal rotation.

In the theoretical treatment of the internal rotation potential referred to above, all the bond angles have been kept constant for all the internal rotation state. However, the theoretical value of the potential barrier may become lower if one considers that the bond angles change to some extent on crossing over the potential barrier. This was found to be actually the case for hexachloroethane $Cl_3C$–$CCl_3$ by Morino and Ikeda[20] who calculated the potential function of internal rotation by use of the Lennard-Jones type of interaction. The calculated value of the barrier height for the molecular configuration with no change of bond angle was found almost twice as much as the minimum value of the barrier height for which the deformation of bond angles was taken into account. This minimum value is in good agreement with the experimental value to be explained in the following.

In such a molecule as hexachloroethane, the rotating groups contain atoms with fairly strong scattering power and accordingly the electron diffraction method may be applied to the experimental determination of the barrier height. In Section 8 we have treated the diffraction pattern of hexachloroethane on the basis of the rigid staggered model, but it can be shown that some of the maxima of the theoretical intensity curve are affected

18. D. R. J. Boyd, R. L. Williams and H. W. Thompson, Nature, **167**, 766 (1951).
19. E. N. Lassettre and L. B. Dean, J. Chem. Phys., **17**, 317 (1949).
20. Y. Morino and S. Ikeda, private communication.

considerably by the torsional motion about the C–C axis. This torsional motion depends on the curvature about the minima of the potential curve which in turn depends on the barrier height, assuming again the cosine function of Eq. (3.5). Such a treatment was made by Morino and Iwasaki who found the best agreement with the experimental results if the barrier height of hexachloroethane was taken as 10 – 15 kcal./mole.[21]

In ethane or hexachloroethane the internal rotation potential has a threefold symmetry and, therefore, the height of the potential barrier has been evaluated rather easily. However, in such a substance as 1,2-dichloroethane there are two kinds of potential barriers in one complete internal rotation which make the height determination considerably difficult. The investigation of Pitzer and Gwinn described in the preceding section is an attempt to estimate the barrier height by making use of the available data of entropy, heat capacity and dipole moment.*

In any case the potential barrier to internal rotation of 1,2-dichloroethane, $ClH_2C$–$CH_2Cl$, is determined by the interactions between Cl and Cl, between H and H and between H and Cl on the different movable groups. Of these the former two interactions can be estimated from the barrier height of ethane and of hexachloroethane as referred to above. The interaction between H and Cl may be known from a study on such a symmetrical molecule as methylchloroform $H_3C$–$CCl_3$, to which a method similar to that used in the case of ethane can be applied. Rubin, Levedahl and Yost[22] found the barrier height of this substance to be 2700 cal./mole from the comparison of the calorimetric entropy with that calculated from molecular constants obtained by spectroscopic measurements, etc. This value of barrier height is almost the same as that in ethane and, therefore, the interaction between H and H and that between Cl and H on different groups seem to be almost equal to each other. This suggests that the lower potential barrier of 1,2-dichloroethane has a height not much different from that of ethane or methylchloroform (about 3 kcal./mole),** and the rotational interconversion takes place at a considerable rate at ordinary temperature, so that the equilibrium between the *trans* and the *gauche* forms is readily

---

* See also H. J. Bernstein, J. Chem. Phys., **17**, 262 (1949).
** It is interesting that almost the same barrier height has been found for methylfluoroform (3000 cal./mole). See H. S. Gutowsky and H. B. Levine, J. Chem. Phys., **18**, 1297 (1950).

---

21. Y. Morino and M. Iwasaki, J. Chem. Phys., **17**, 216 (1949).
22. J. R. Rubin, B. H. Levedahl and D. M. Yost, J. Am. Chem. Soc., **66**, 279 (1944).

reached. It is evident that the other potential barrier is much higher than the one referred to above, since in that case one chlorine atom on one movable group eclipses the other chlorine atom on the other group as viewed along the C–C axis.

If the potential barrier is sufficiently high, so that the rotational isomers can be separated from each other, we may measure the rate of transformation in order to determine the barrier height. For example, some derivatives of diphenyl which we have discussed in Section 1 have been resolved into their $d$ and $l$ forms. The racemization of these optically active compounds has been investigated by Cagle and Eyring by the use of the absolute reaction rate theory.[23] They obtained values for the heat of activation ranging from 20 kcal./mole (2-nitro-6-carboxy-6'-ethoxydiphenyl) to 46 kcal./mole (2,2'-diamino-6,6'-dimethyl-diphenyl).

In this case the heat of activation may contain, in the first approximation, two terms. One is the work necessary to distort the bonds so that the molecule can be brought into a coplanar configuration. The other term is the resonance energy which results from this configuration. Combining suitably the experimental data obtained for different substances, Cagle and Eyring were able to estimate the resonance energy as 0.5 kcal./mole.

Before concluding the discussion of this section it is necessary to make a remark on an uncertain frequency of ethane in its contribution to thermal properties and on the equilibrium configuration of this molecule. As referred to above, the contribution of the torsional frequency to heat capacity can at once be obtained by the difference method, if we know all the other frequencies of the ethane molecule. However, one of them has not been observed as a fundamental frequency and has been determined from combination tones. It is, therefore, no wonder that there was once some ambiguity as to the frequency value of this vibration.

If we assign a frequency of 750 cm.$^{-1}$ to this vibration according to Eucken et al., we are led to the low value of barrier height as first suggested by them, while if we adopt a frequency value of about 1100 cm.$^{-1}$ as first proposed by Howard,[24] we can arrive at a barrier height of 3 kcal./mole. The latter value (1155 cm$^{-1}$. of Table 3.2) is now generally accepted from the normal vibration calculations made by several authors as explained below.

23.  F. Wm. Cagle and H. Eyring, J. Am. Chem. Soc., **73**, 5628 (1951).
24.  J. B. Howard, J. Chem. Phys., **5**, 442, 451 (1937).

Evidently, if the heat capacity is measured at lower temperatures, the contribution of this inactive frequency becomes very small so that the ambiguity of the frequency value does not come into question. As referred to above, such a measurement has been made by Kistiakowsky. Lacher and Stitt who found a barrier height of 3 kcal./mole.[14]

TABLE 3.1

Vibrations of the staggered configurations of ethane

| Class | Symmetric | Anti-symmetric | Raman | Infrared | Number of vibrations | Type of vibrations |
|-------|-----------|----------------|-------|----------|----------------------|--------------------|
| $A_{1g}$ | $C_3, \sigma_v, i$ | — | active | inactive | 3 | CH stretching<br>$CH_3$ deformation<br>CC stretching |
| $A_{1u}$ | $C_3$ | $\sigma_v, i$ | inactive | inactive | 1 | Torsion |
| $A_{2g}$ | $C_3, i$ | $\sigma_v$ | inactive | inactive | 0 | |
| $A_{2u}$ | $C_3, \sigma_v$ | $i$ | inactive | active | 2 | CH stretching<br>$CH_3$ deformation |
| $E_g$ | $i$ | — | active | inactive | 3 | CH stretching<br>$CH_3$ deformation<br>Bending |
| $E_u$ | — | $i$ | inactive | active | 3 | CH stretching<br>$CH_3$ deformation<br>Bending |

Although we have not yet shown above the direct evidence that the stable configuration of ethane is the staggered model $D_{3d}$, all the experimental results concerning the ethane derivatives referred to in this chapter and the previous one are in favor of this configuration. We have, therefore, one three-fold axis $C_3$, three planes of symmetry $\sigma_v$ and a center of symmetry $i$ and the eighteen normal vibrations can be divided into classes as shown in Table 3.1, where the selection rules for the Raman effect and the infrared absorption and the types of vibrations are also tabulated. For example the three types of vibrations (CH stretching, $CH_3$ deformation and CC stretching vibrations) belonging to the class $A_{1g}$ are symmetric to the axis, the plane and the center of symmetry: in other words they are all totally symmetric vibrations and are active in the Raman effect but forbidden in the infrared absorption.

By 1938, Crawford, Avery, and Linnett were able to give a complete summary of the twelve distinct fundamental frequencies,[25] although the spectroscopic data available at that time were insufficient to determine the actual equilibrium configuration of the ethane molecule. In 1939 Stitt measured the infrared and Raman spectra of heavy ethane $C_2D_6$ and determined a set of potential constants.[26] Recently, the high resolution measurements by Smith considerably improved and extended the infrared data for ethane.[27] The observed combination bands gave direct evidence for the $D_{3d}$ equilibrium configuration. Additional high resolution measure-

TABLE 3.2

Assignment of vibrational frequencies of ethane and heavy ethane*

| Class | Type of vibration | $C_2H_6$ | $C_2D_6$ |
|-------|-------------------|----------|----------|
| $A_{1g}$ | CH stretching | 2899 | 2083 |
|          | $CH_3$ deformation | 1375 | 1158 |
|          | CC stretching | 993 | 852 |
| $A_{1u}$ | Torsion | 275 | 200 |
| $A_{2u}$ | CH stretching | 2954 | 2111 |
|          | $CH_3$ deformation | 1379 | 1072 |
| $E_g$ | CH stretching | 2963 | 2225 |
|       | $CH_3$ deformation | 1460 | 1055 |
|       | Bending | 1155 | 970 |
| $E_u$ | CH stretching | 2994 | 2236 |
|       | $CH_3$ deformation | 1486 | 1102 |
|       | Bending | 821 | 601 |

* See G. Herzberg, Infrared and Raman Spectra of Polyatomic Molecules, D. Van Nostrand Company, INC., New York (1945).

25. B. L. Crawford, W. H. Avery and J. W. Linnett, J. Chem. Phys., 6, 682 (1938).
26. F. Stitt, J. Chem. Phys., 7, 297 (1939).
27. L. G. Smith, J. Chem. Phys., 17, 139 (1949).

ments on the infrared spectrum of heavy ethane were made by Hansen and Dennison[28] in order to reexamine the spectroscopic data on light and heavy ethane and to determine the potential constants as accurately as possible. From the resolution of the fine structure of the symmetric parallel band, the dimensions of the ethane molecule have been obtained as follows: C–C distance 1.543Å, C–H distance 1.102Å, H–C–C angle 109°37′ and H–C–H angle 109°19′.

In Table 3.2 are shown the assignments of the fundamental frequencies of $C_2H_6$ and $C_2D_6$ to the types of vibration given in Table 3.1.

## 11. Internal Hydrogen Bond.
### Entropy Difference between the Rotational Isomers

As stated in the preceding sections the molecules of 1,2-dihalogenoethanes exist in the *trans* and the *gauche* forms. This is true for the molecules of the general type of $XH_2C–CH_2X$ or $XH_2C–CH_2Y$ where X and Y denote Cl, Br, I, $CH_3$, OH, etc. In most of the molecules the *trans* form is more stable than the *gauche* form. However, in such a molecule as ethylene chlorhydrin $ClH_2C–CH_2OH$, the *gauche* form may become more stable than the *trans* form, because the internal hydrogen bond is formed in the former, but not in the latter.

We shall discuss this problem in this section, before we proceed to the general discussion of the nature of the hindering potential to be made in the following section.

Mizushima, et al.[29] observed the Raman and infrared spectra of this substance and concluded that the molecules take both the *trans* and the *gauche* forms in the gaseous and the liquid states, while they take only the *gauche* form in the solid state in contradistinction to the case of 1,2-dihalogeno-ethanes in which only the *trans* form persists in the solid state. They also observed the temperature-dependence of the infrared absorption in the gaseous state, from which they concluded that the *trans* form has an excess of energy of 0.95 kcal./mole as compared with the *gauche* form.

As stated above this is considered to be due to the internal hydrogen bond formed between the chlorine atom and hydroxyl group. If this is

28. G. E. Hansen and D. M. Dennison, J. Chem. Phys., **20**, 313 (1952). See also the electron diffraction data recently obtained by K. Hedberg and V. Schomaker, J. Am. Chem. Soc., **73**, 1482 (1951).

29. S. Mizushima, T. Shimanouchi, T. Miyazawa, K. Abe and M. Yasumi, J. Chem. Phys., **19**, 1477 (1951).

actually the case, the *gauche* molecule is restricted considerably in its internal motion and we may expect a large entropy difference between the *trans* and *gauche* molecules. In order to determine the entropy difference between such rotational isomers, Mizushima, Shimanouchi, Kuratani and Miyazawa presented an optical method which we shall now explain.[30]

The ratio of the *trans* and the *gauche* molecules in equilibrium can be expressed as

$$\frac{N_t}{N_g} = \frac{1}{2} e^{\frac{\Delta S}{R}} e^{-\frac{\Delta E}{RT}} \tag{3.7}$$

where $\Delta S = S_t - S_g$ and $\Delta E = E_t - E_g$ are, respectively, the entropy difference and the energy difference between the *trans* and *gauche* isomers.

Let us consider the three bands of the infrared absorption $T$, $G$ and $S$ of which $T$ is the absorption band characteristic of the *trans* molecule and $G$ that of the *gauche* molecule and $S$ the standard band common to the *trans* and the *gauche* molecules in frequency and in molecular absorption coefficient.

From Eq. (3.7) we have

$$\frac{N_t}{N_t + N_g} = \frac{1}{1 + 2 e^{-\frac{\Delta S}{R} + \frac{\Delta E}{RT}}}$$

$$\frac{N_g}{N_t + N_g} = \frac{1}{1 + \frac{1}{2} e^{\frac{\Delta S}{R} - \frac{\Delta E}{RT}}}$$

Let $\varkappa$ be the molar absorption coefficient, and $D$ be the optical density $(D = \ln I_0/I)$. Then we have

$$\frac{D_T}{D_S} = \frac{\varkappa_t}{\varkappa_s} \frac{N_t}{N_t + N_g} = \frac{\varkappa_t}{\varkappa_s} \frac{1}{1 + 2 e^{-\frac{\Delta S}{R} + \frac{\Delta E}{RT}}}$$

$$\frac{D_G}{D_S} = \frac{\varkappa_g}{\varkappa_s} \frac{N_g}{N_t + N_g} = \frac{\varkappa_g}{\varkappa_s} \frac{1}{1 + \frac{1}{2} e^{\frac{\Delta S}{R} - \frac{\Delta E}{RT}}}$$

$$2 e^{\frac{\Delta E}{RT}} e^{-\frac{\Delta S}{R}} - \frac{\varkappa_t}{\varkappa_s} \frac{D_S}{D_T} + 1 = 0 \tag{3.8}$$

$$\frac{1}{2} e^{-\frac{\Delta E}{RT}} e^{\frac{\Delta S}{R}} - \frac{\varkappa_g}{\varkappa_s} \frac{D_S}{D_G} + 1 = 0 \tag{3.9}$$

30. S. Mizushima, T. Shimanouchi, K. Kuratani and T. Miyazawa, J. Am. Chem. Soc., **74**, 1378 (1952).

The energy difference $\varDelta E$ has been obtained as 0.95 kcal./mole as stated above and therefore, if we measure $D_S/D_T$ or $D_S/D_G$ at different temperatures, we can evaluate $\varDelta S$. Let $T$ be the band at 760 cm.$^{-1}$, $G$ that at 669 cm.$^{-1}$ and $S$ that at 2890 cm.$^{-1}$ we have

$$\varDelta S = 3.7 \pm 0.4 \text{ e. u.}$$

i. e. the entropy of the *trans* isomer is larger by 3.7 e. u. than that of the *gauche* isomer.

It will easily be seen from the values of the moments of inertia and normal frequencies of the *trans* and the *gauche* molecules that this entropy difference arises mainly from the difference in the internal rotation about the $C - C$ and $C - O$ axes.*

In the *trans* molecule the H-atom of the OH-group has three stable positions with regard to the internal rotation about $C - O$ axis and the energy difference among these positions should be very small.   Similarly, in the *gauche* molecule there are three such stable positions of which, however, one has lower energy than the other two owing to the internal hydrogen bond. Therefore, the contributions of these two positions would be small in the *gauche* molecule and the entropy of this isomer would become

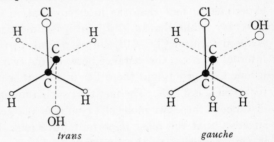

Fig. 3.1.  Stable configurations of $ClH_2C-CH_2OH$ as viewed along the C–C axis.

considerably smaller than that of the *trans* isomer.  Moreover, due to this internal hydrogen bond the frequencies of the torsional oscillation will become higher in the *gauche* molecule than in the *trans* molecule.  As these frequencies are of the order of magnitude of 100 cm.$^{-1}$, such changes in their values will make the entropy of the *gauche* isomer considerably smaller than that of the *trans* isomer.

Thus we have good reason to believe that the main part of the entropy difference between the rotational isomers of ethylene chlorhydrin arises from the difference in internal rotation which is due to the fact that only one of the isomers has an internal hydrogen bond.

---

* The calculation shows that the entropy difference arising from the overall rotation and normal vibrations is less than one tenth of the observed value. (See ref. 30.)

## 12. The Nature of the Hindering Potential

In the preceding sections we have described various experimental results concerning the internal rotation of ethane derivatives. We are now in a position to make at least a qualitative discussion of the nature of the hindering potential (Mizushima, Morino and Shimanouchi[31]). In Table 3.3 are summarized the stable configurations of the rotational isomers so far discussed, together with some of those to be considered in the following sections.

From the data shown in Table 3.3 we can conclude that the most important force determining the stable configurations of rotational isomers is the steric repulsion between two groups rotating against each other about a single bond as axis. For example, the fact that the molecules of the type of $XH_2C-CH_2Y$ have rotational isomers of the same configurations (i. e. the *trans* and the *gauche* forms) irrespective of the dipole moment values of C–X and C–Y bonds reveals that the steric repulsion is far more important than the electrostatic force. This conclusion is also compatible with the experimental fact that the internal rotation of $Cl_3C-CCl_3$ is hindered considerably, while that of $Cl_3Si-SiCl_3$ is almost free. (See Section 16.) In the eclipsed form of $Cl_3C-CCl_3$ corresponding to the potential maximum of the internal rotation, the distance between the two chlorine atoms, one contained in one $Cl_3C$-group and the other in the other group, is 2.72Å, while that of $Cl_3Si-SiCl_3$ is 3.20Å. The fact that such a minor difference in the interatomic distance in these two molecules results in a large difference in internal rotation suggests that the most important force must be the steric repulsion which is very sensitive to the variation of interatomic distance.

Concerning the steric repulsion in $Cl_3C-CCl_3$ we should like to note that in the staggered form (i. e. the stable form) the distance between two nearest chlorine atoms on different movable groups is not much different from the sum of the van der Waals radii of two chlorine atoms, while in the eclipsed form this distance becomes 2.72Å at which distance there must be a considerable repulsion. This is the most important reason why the eclipsed form corresponds to the potential maximum of internal rotation.

In the stable configurations, therefore, the repulsive potential tends to assume as low a value as possible and the repulsive force may become of the same order of magnitude as electrostatic force, hydrogen bond, etc. This may be seen, for example, from the value of the energy difference between the rotational isomers of $Br(CH_3)_2C-C(CH_3)_2Br$ ($\Delta E = 1.6$ kcal./mole) which is even greater than that of $BrH_2C-CH_2Br$ ($\Delta E = 1.4$ kcal./mole).

---

31. S. Mizushima, Y. Morino and T. Shimanouchi, J. Phys. Chem., **56**, 324 (1952).

Since a $CH_3$-group has almost the same van der Waals radius as a Br-atom, the $\Delta E$ of $Br(CH_3)_2C$–$C(CH_3)_2Br$ would not be so large, if only the steric repulsion contributed to the energy difference between the rotational isomers. In such a case we must, therefore, consider that there is an important contribution of the electrostatic force to $\Delta E$.

TABLE 3.3

Stable configurations of rotational isomers.[31] (Figures in parentheses attached to the less stable isomer denote the energy difference in kcal./mole).

| Molecule | Solid | Liquid | Gas | $CCl_4$-solution |
|---|---|---|---|---|
| $ClH_2C$–$CH_2Cl$ | *trans* | *trans* gauche (ca. 0) | *trans* gauche (1.2) | *trans* gauche |
| $BrH_2C$–$CH_2Br$ | *trans* | *trans* gauche (0.7) | *trans* gauche (1.4) | *trans* gauche |
| $ClH_2C$–$CH_2Br$ | *trans* | *trans* gauche (0.4) | | *trans* gauche |
| $ClH_2C$–$CH_2I$ | *trans* | *trans* gauche | | *trans* gauche |
| $CH_3CH_2$–$CH_2CH_3$ | *trans* | *trans* gauche (0.8) | *trans* gauche | |
| $CH_3CH_2$–$CH_2Cl$ | *trans* | *trans* gauche | *trans* gauche | |
| $HOH_2C$–$CH_2Cl$ | *gauche* | *trans* gauche | *trans* (0.95) gauche | |
| $ClH_2C$–$COCl$ | *trans* | *trans* gauche | *trans* gauche | |
| $BrH_2C$–$COCl$ | *trans* | *trans* gauche | *trans* gauche (1.0) | |
| $BrH_2C$–$COBr$ | *trans* | *trans* gauche | *trans* gauche (1.9) | |
| $ClH_2C$–$COCH_3$ | *trans* | *trans* gauche | *trans* gauche | |
| $ClOC$–$COCl$ | *trans* | *trans* cis | | |
| $Br(CH_3)_2C$–$C(CH_3)_2Br$ | | | | *trans* gauche (1.6) |
| $Cl(CH_3)_2C$–$C(CH_3)_2Cl$ | | | | *trans* gauche (1.0) |
| $Cl_3C$–$CCl_3$ | staggered | | | |
| $Cl_3Si$–$SiCl_3$ | | almost free rotation | almost free rotation | |

It is, however, evident that even in the stable configurations we cannot neglect the steric repulsion. This will be seen, for example, from the difference in $\Delta E$ between $ClH_2C-CH_2Cl$ ($\Delta E = 1.2$ kcal./mole) and $BrH_2C-CH_2Br$ ($\Delta E = 1.4$ kcal./mole) shown in Table 3.3. The contribution of electrostatic force to $\Delta E$ in the dichloride must be greater than that in the dibromide, since the bond moment of C–Cl is almost equal to that of C–Br and the induction effect in the dichloride is smaller than that in the dibromide.* Therefore, the fact that $\Delta E$ of $ClH_2C-CH_2Cl$ is smaller than that of $BrH_2C-CH_2Br$ shows that there is a considerable contribution of the steric repulsion to the $\Delta E$ of such rotational isomers.

It would be very interesting to determine the ratio of the steric part to the electrostatic part of the hindering potential. The value of $\Delta E$ (0.8 kcal./mole) of $n$-butane may tell us the approximate magnitude of the steric part of $\Delta E$ of $BrH_2C-CH_2Br$, since $n$-butane is a non-polar substance and the van der Waals radius of $CH_3$-group is almost equal to that of Br-atom as stated above.

In such a substance as $ClH_2C-CH_2OH$ an internal hydrogen bond is formed between the two rotating groups. As stated in the preceding section this bond lowers the energy of the *gauche* form to make it more stable than the *trans* form which otherwise would be always the lower energy form in the case of molecules of the type of $XH_2C-CH_2Y$ (see Table 3.3). However, the internal hydrogen bond is not so strong as to shift the *gauche* position up to the *cis* position.

In the case where the internal rotation axis acquires double bond character to a considerable extent, the situation may become quite different from what we have described above. Among the substances shown in Table 3.3 ClOC–COCl provides us with such an example. For this substance the *trans* form is found to be the stable configuration, which would not be the case if steric repulsion were the predominant factor in determining the stable configuration. The stability of the *trans* form in this case is due on one hand to the single bond-double bond resonance of the structure of

$$O{=}C{-}C{=}O$$

and on the other hand to the fact that the steric repulsion between two Cl-atoms is much greater than that between a Cl- and an O-atom.

---

* This would be seen from the difference in the moment value $\mu_g$ of the *gauche* molecule between the dichloride and the dibromide. $\mu_g$ of $ClH_2C-CH_2Cl$ is $2.55D$, while $\mu_g$ of $BrH_2C-CH_2Br$ is $2.0D$.

So far we have mainly discussed the molecular configurations and the energy differences in the gaseous state. We have, however, often seen that considerable changes of the equilibrium ratio of different configurations occur with a change of state. For example, for $ClH_2C$–$CH_2Cl$ the *gauche* molecule becomes more abundant in the liquid state than in the gaseous state, while in the solid state the *gauche* molecule disappears almost completely. Such a change in equilibrium ratio must be explained in terms of the intermolecular forces as has been already discussed in the preceding sections. However, as we can see from Table **3.3**, the intermolecular force is not strong enough to change greatly the positions of maxima and minima of the hindering potential which is determined mainly by steric repulsion in case where there is no single bond-double bond resonance of an appreciable amount.

## Summary of Chapter III

The spectroscopic and electric measurements described in the preceding chapter provide us with almost all the molecular data necessary for the calculation of thermodynamic functions of 1,2-dihalogenoethanes. We are lacking in only one molecular constant: this is the torsional frequency about the C – C bond of the *trans* molecule.

The contribution of this torsional motion to entropy can be obtained by subtracting from the thermal entropy, the molecular entropy calculated without taking into account the torsional motion. From this difference the torsional frequency can be calculated, assuming this motion to be harmonic. The value is found to be about 100 cm.$^{-1}$, which is not much different from those of the lowest Raman frequencies (125 cm.$^{-1}$ for $ClH_2C$–$CH_2Cl$ and 91 cm.$^{-1}$ for $BrH_2C$–$CH_2Br$) assigned to the torsional motion of the *gauche* form. The corresponding *trans* frequency is inactive in the Raman effect. If we consider this frequency to be about 100 cm.$^{-1}$, (which is quite reasonable), we can explain the entropy value in exactly the same way as we have explained the spectroscopic and electric data.

In such symmetrical molecules as $H_3C$–$CH_3$, $H_3C$–$CCl_3$, and $Cl_3C$–$CCl_3$, which have three equivalent potential minima in one internal rotation, we have only one kind of torsional motion, which makes the theoretical treatment quite easy. In this case we can calculate the contribution of this motion to entropy by subtracting from the thermal entropy the molecular entropy without the contribution of torsional motion. From this we can also calculate the height of the potential barrier to internal rotation, assuming the potential function to be of the sine form.

The barrier height was found to be 2750 cal./mole for $H_3C$–$CH_3$ and 2700 cal./mole for $H_3C$–$CCl_3$. From these values we may estimate the height of the lower barriers of $ClH_2C$–$CH_2Cl$ to be about 3000 cal./mole, but the other barrier which lies at the *cis* position should be much higher than these barriers, since the two chlorine atoms come at the shortest distance just as in the eclipsed form of $Cl_3C$–$CCl_3$, for which the barrier height was found to be 10 – 15 kcal./mole.

As to the nature of the hindering potential, the most important element has been shown to be the steric repulsion between the movable groups. The electrostatic interaction which was once thought to be most important plays a less significant role as compared with the steric repulsion. Therefore, the similar steric potential always yields the similar stable configurations, irrespective of the electrostatic interaction. For example, $ClH_2C$–$CH_2Cl$ has the stable configurations similar to that of $H_3CH_2C$–$CH_2CH_3$, although the former has highly polar movable groups while the latter nonpolar groups.

In the stable configurations, in which the repulsive potential tends to take as low a value as possible, the steric repulsion may become of the same order of magnitude as electrostatic force. This means that the electrostatic potential may affect considerably the energy difference between the different stable configurations. This was shown to be actually the case from the comparison of the values of energy difference between the polar and nonpolar substances such as $BrH_2C$–$CH_2Br$ and $H_3CH_2C$–$CH_2CH_3$, both of which are considered to have almost the same potential, so far as the steric repulsion is concerned.

If there is a hydrogen-bond between the movable groups, this affects the energy difference more conspicuously than the usual electrostatic force, although the main feature of the potential curve continues to be determined by the steric repulsion. In this case we find a large difference in entropy between the rotational isomers, which can be determined from the intensity measurement of the infrared absorption band.

An exceptional case, where the steric repulsion does not play the predominant role in the internal rotation potential, is that where the rotation axis acquires double bond character to a considerable amount. In this case we may have a stable configuration which does not correspond to the minimum of the steric repulsive potential.

# CHAPTER IV

# Internal Rotation in Other Simple Molecules

## 13. Other Simple Molecules with C-C Axes

In the preceding two chapters we have confined ourselves to the discussion on simple ethane derivatives in order to make straightforward explanation of the problems of internal rotation. There are some other simple molecules with C – C axes of internal rotation which remain to be explained.

Let us first discuss the rotational isomerism of $n$-propyl chloride $ClH_2C–CH_2CH_3$ which has a skeleton similar to that of 1,2-dihalogenoethanes but is quite different from them from an electrostatic point of view. According to what has been stated with regard to the nature of hindering potential (see Section 12), the stable configurations of this substance should be similar to those of 1,2-dihalogenoethanes: i. e. we expect that the substance exists in the *trans* and the *gauche* forms in the liquid and gaseous states and only one of these forms persists in the solid state. This has been shown to be actually the case by the measurements of the Raman effect and infrared absorption.[1, 2] The configuration which persists in the solid state has been shown to be the *trans* form.[2]

The measurement of the temperature-dependence of the absorption intensity shows that the energy difference between the *trans* and the *gauche* molecules is small in the gaseous state.[2] This has also been found to be the case in the liquid state. This is quite understandable, since in this case the *trans* and the *gauche* forms have moment values equal to each other and accordingly, both of them are stabilized in the liquid state to the same extent.

Ethylene chlorhydrin explained in Section 11 and $n$-butane to be discussed in Chapter V are also molecules with skeletons similar to 1,2-dihalogenoethanes and both have stable configurations similar to those of 1,2-dihalogenoethanes.

---

1. S. Mizushima, Y. Morino and S. Nakamura, Sci. Pap. Inst. Phys. Chem. Res. Tokyo, **37**, 205 (1940).

2. I. Ichishima, C. Komaki, T. Miyazawa, K. Kuratani and S. Mizushima, unpublished.

As explained in Section 12, the most important force restricting the internal rotation is the steric repulsion. However, this force decreases very rapidly with distance and accordingly in such a molecule as dimethyl acetylene, $CH_3C \equiv CCH_3$, the two methyl groups can rotate quite freely as has been shown by the thermal measurement referred to in Section 11. We can expect the same situation for dichlorobutyne $ClH_2C-C \equiv C-CH_2Cl$, in which the two $ClH_2C$-groups are separated by a distance far enough from each other. However, there is a difference in that this molecule has a fairly large electrostatic potential which may become more significant than the steric potential at a large distance. If this is the case, there will be a potential barrier of an appreciable height, which may be determined by the dipole measurement. In conformity with this expectation, Morino, Miyagawa and Wada[3] found a small temperature-dependence of the moment value from which they calculated the barrier height as 0.75 kcal./mole, assuming the simple type of potential function of Eq. (1.1). They could also calculate approximately this barrier height from the bond moments of C–Cl and C–H, in order to show that the hindering potential is in this case essentially of an electrostatic nature. In contrast to the case of 1,2-dichloroethane there is no anomaly for this substance in benzene solution. In other words, the apparent moment found in benzene solution is just the magnitude to be expected from the dielectric constant of benzene. Therefore, the fact that we found in 1,2-dichloroethane an unexpectedly large moment value in benzene solution is due to some special circumstance which favors an interaction between the solvent and solute molecules. This interaction does not exist appreciably when two $ClH_2C$-groups are separated from each other far apart as in dichlorobutyne.

For 1,2-dichloropropane, $ClH_2C-CHClCH_3$, the three stable configurations shown in Fig. 4.1 are conceivably possible. However, in g' form (gauche-2) a chlorine atom is situated at the gauche position both to another chlorine atom and to the methyl group and accordingly, this form has much higher energy than the other two. This means that at ordinary temperatures this form can be neglected and the substance can be treated as an isomeric mixture of t and g forms. The result of the Raman measurement of Kahovec and Wagner[4] is compatible with this view.

Oriani and Smyth[5] have measured the variation of the gas phase dipole moment with temperature, the result of which has been discussed by Morino,

3.  Y. Morino, I. Miyagawa and A. Wada, J. Chem. Phys., 20, 1976 (1952).
4.  L. Kahovec and J. Wagner, Z. phys. Chem. (B) 47, 48 (1940).
5.  R. A. Oriani and C. P. Smyth, J. Chem. Phys., 17, 1174 (1949).

Miyagawa and Haga[6] in the same way as in the case of gaseous 1,2-dichloro-ethane explained in Section 6. They found the energy difference between g and t forms to be 1.0 kcal./mole, which is smaller than that of 1,2-dichloro-ethane (1,2 kcal./mole). This is quite understandable, because the steric repulsion between the chlorine atom and methyl group makes the t form of 1,2-dichloropropane less stable than the corresponding form of 1,2-dichloro-ethane, while there would be no appreciable difference in steric potential between the g forms of these two substances.

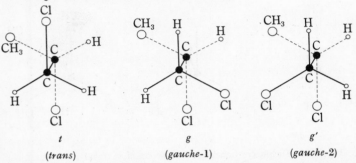

| t | g | g' |
|---|---|---|
| (*trans*) | (*gauche*-1) | (*gauche*-2) |

Fig. 4.1. The conceivable stable configurations of 1,2-dichloropropane.

If we replace the methyl group of 1,2-dichloropropane by a chlorine atom, we obtain 1,1,2-trichloroethane. In this case the t and g forms degenerate into one form and accordingly, the assumption of the coexistence of t and g forms of 1,2-dichloropropane (or neglect of g' form) leads to the assumption of almost one molecular form for 1,1,2-trichloroethane. This is just what we have concluded from the gas-phase moment value of this substance (Section 6). We have seen that the other conceivable form which corresponds to the g' form of 1,2-dichloropropane has much higher energy than the stable form referred to above.

A molecule such as secondary butyl alcohol $CH_3CH_2-C^*HOHCH_3$, has an asymmetric carbon atom (marked with the asterisk) and exists, therefore, as D- and L-forms. Due to the internal rotation about the C – C bond, each of the optically active forms may have three stable configurations shown in Fig. 4.2 for one of the optical isomers.

According to Kirkwood[7] we can expect that each of the rotational isomers will rotate the plane of polarization of incident light to a different degree.

6. Y. Morino, I. Miyagawa and T. Haga, J. Chem. Phys., 19, 791 (1951).
7. J. G. Kirkwood, J. Chem. Phys., 5, 479 (1937).

Since the angle of rotation is proportional to the number of active molecules present, the rotation of the equilibrium mixture of the three rotational isomers will depend on the equilibrium concentrations of the three forms. Therefore, the measurement of the specific rotation will give us information about the equilibrium concentration of the three rotational isomers referred to above. This is another interesting method for the experimental study of rotational isomerism.

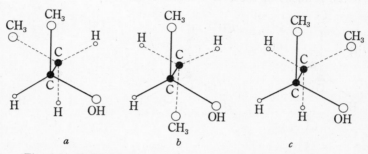

Fig. 4.2.  Stable configurations of secondary butyl alcohol.

Such a kind of measurement was first made by Bernstein and Pedersen[8] who found the concentrations of the three rotational isomers of D-secondary butyl alcohol as 42.3 percent, 42.3 percent and 15.3 percent, assuming that configurations $a$ and $b$ have the same energy and that $c$ is less stable than either $a$ or $b$. (Their measurement was made in dilute solutions of cyclohexane at different temperatures. After corrections for solvent effect and change of refractive index with temperature had been made, there remained a definite temperature-dependence of the rotation.)

So far we have discussed the internal rotation about the $C - C$ axis in which both the carbon atoms are combined with other atoms through single bonds. Let us next discuss the internal rotation about the $-CH_2-C-$ axis

$$\overset{\|}{O}$$

in order to see the effect of the oxygen atom combined by a double bond. This is important for the structural chemistry of proteins, because the $-CH_2-CO-$ axis is one of the axes of internal rotation of a polypeptide chain.

The Raman and the infrared spectra of chloroacetyl chloride $ClH_2C-COCl$, bromoacetyl chloride $BrH_2C-COCl$ and bromoacetyl bromide $BrH_2C-COBr$

---

8.  H. J. Bernstein and E. E. Pedersen,  J. Chem. Phys., **17**, 885 (1949).

have been measured by Nakagawa et al.[9]  From the experimental results it has been concluded that there are two rotational isomers in the liquid and gaseous states and only one of them persists in the solid state.  From the temperature-dependence of the intensity of the infrared absorption in the gaseous state the energy difference between these two isomers has been found to be 1.0 kcal./mole for bromoacetyl chloride and 1.9 kcal./mole for bromoacetyl bromide.  These authors also calculated the normal vibrations and the product rule for various configurations of rotational isomers and showed that the more stable form is the *trans* form (or nearly this one) with regard to the two halogen atoms.  The less stable form is considered to differ from the *trans* form through an azimuthal angle of about 150° (see Fig. 4.3).

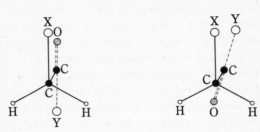

Fig. 4.3.  Stable configurations of XH$_2$C–COY (X and Y denoting halogen atoms).

Similar investigation has been made by Mizushima et al.[10] on chloroacetone, ClH$_2$C–COCH$_3$, which has a similar skeletal structure to the substances just stated above, but is different from them in its electrostatic potential.  The Raman, infrared and dipole measurements, together with the calculation of normal vibrations and of the product rule, showed that chloroacetone exists in two molecular forms similar to those of ClH$_2$C–COCl in the liquid and gaseous states and in one form (the *trans* form, or the more polar form) in the solid state.

It is worthy of note that the relative intensity of two absorption bands at 728 and 764 cm.$^{-1}$ is reversed, when we go from the liquid state to the gaseous state.  In the liquid state the band at 764 cm.$^{-1}$ assigned to the more polar form is stronger than the band at 728 cm.$^{-1}$ assigned to the less polar form, while in the gaseous state the band at 764 cm.$^{-1}$ becomes much weaker than that at 728 cm.$^{-1}$.  In other words there is a conspicuous change of the isomeric ratio on condensation.  Of the two bands referred to above, only the 764 cm.$^{-1}$ band persists in the solid state.

9.  I. Nakagawa, I. Ichishima, K. Kuratani, T. Miyazawa, T. Shimanouchi and S. Mizushima,  J. Chem. Phys., **20**, 1720 (1952).

10.  S. Mizushima, T. Shimanouchi, T. Miyazawa, I. Ichishima, K. Kuratani, I. Nakagawa and N. Shido,  J. Chem. Phys., **21**, 815 (1953).

We have often stated that the hindering potential is determined first by the steric repulsion and then by the electrostatic interaction. If, therefore, we assume that the steric repulsion between chlorine and oxygen atoms is much less than that between two chlorine atoms or between chlorine atom and methyl group, both of the molecules of $ClH_2C-COCl$ and $ClH_2C-COCH_3$ tend to take the *trans* form with respect to two chlorine atoms or chlorine atom and methyl group. However, the electrostatic interaction favors the other configuration (with the azimuthal angle of about 150°) in $ClH_2C-COCH_3$, since in this configuration the two large bond moments at C–Cl and C=O are almost in opposite directions. In the case of $ClH_2C-COCl$ we have another C–Cl bond which makes the difference in moment value between the two molecular configurations much less significant than in the case of $ClH_2C-COCH_3$. From this we can understand the experimental result referred to above, viz. that on condensation the isomeric ratio of $ClH_2C-COCH_3$ changes much more markedly than that of $ClH_2C-COCl$, since the stabilization of a molecular configuration in the liquid state increases with increasing polarity.

As stated above the most interesting difference of the internal rotation of the molecules with C = O groups from that of 1,2-dihalogenoethanes, etc. is as follows: In the dihalogenoethanes the stable forms as determined by the steric repulsion are also stable from the electrostatic point of view. However, in some molecules with C = O groups, the configuration which is electrostatically more stable becomes less stable from the steric repulsive potential. Evidently this force is more important than the electrostatic force in determining the stable molecular configurations. In conclusion, in the case of the molecules with C = O groups we have also two stable configurations similar to those of the 1,2-dihalogenoethanes; one of them is the *trans* form and the other is the *gauche* form with larger azimuthal angle.

If we have two C = O groups attached directly to the rotation axis, this partially acquires the double bond character.* In this case the molecule tends to take the planar configuration due to the electronic structure of the axis which may affect the internal rotation potential more than the steric repulsion. This would otherwise be the most important force in determining the potential curve, as stated in Section 12.

With regard to such molecules, diacetyl $CH_3OC-COCH_3$, oxamide $H_2NOC-CONH_2$, oxalyl chloride $ClOC-COCl$, and methyl oxalate $CH_3OOC$

---

* This is due to the contribution of a resonance structure which contains a double bond in the rotation axis (C = C). The existence of such a resonance has been proved experimentally in our laboratory by ultraviolet absorption. (Ref. 11.) See also Section 20.

–COOCH$_3$ have been studied in our laboratory[11] by the measurements of the ultraviolet, infrared and Raman spectra. It has been concluded from the results of these measurements that the most stable configuration for all these molecules is the *trans* form and the crystals of these substances contain only this molecular configuration. This is quite reasonable, since of the two

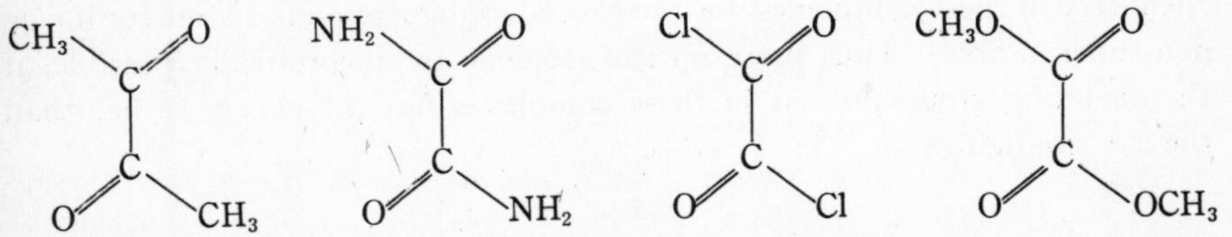

Fig. 4.4. The most stable configurations of diacetyl, oxamide, oxalyl chloride and methyl oxalate.

conceivable planar configurations, (the *trans* and the *cis* forms), the *trans* form would be more stable from the steric and electrostatic points of view. The liquid spectrum of oxalyl chloride contains more Raman lines and absorption peaks than the solid spectrum. This may be explained either by the coexistence of the *cis* form or by considering a broad potential minimum about the *trans* position.

Before concluding the description of the internal rotation about the C – C axis, we would like to refer to a result of the infrared measurement which bears an important relation to the chemistry of metal coordination compounds.

Quagliano and Mizushima[11a] has shown that 1,2-dithiocyanatoethane, NCS·H$_2$C–CH$_2$·SCN, exists in the *trans* and the *gauche* configurations in chloroform solutions but it exists only in the *trans* configuration in the solid state. The infrared spectrum of the complex [PtCl$_2$(NCS·H$_2$C–CH$_2$·SCN)] has been found quite similar to that of the *gauche* configuration of 1,2-di-thiocyanatoethane, but quite different from that of the *trans* configuration. Therefore, the configuration of this chelate ligand in the coordination complex

11. T. Miyazawa, E. Tsunetomi, M. Katayama, H. Baba, K. Kuratani, T. Shima-nouchi and S. Mizushima, unpublished. See also K. W. F. Kohlrausch and H. Wittek, Z. phys. Chem., (B) 48, 177 (1941), J. S. Ziomek, F. F. Cleveland and A. G. Meister, J. Chem. Phys., 17, 669 (1949) and B. D. Saksena and R. E. Kagarise, J. Chem. Phys., 19, 987 (1951), R. E. Kagarise and D. H. Rank, J. Chem. Phys., 19, 1613 (1953), R. E. Kagarise, J. Chem. Phys., 19, 1615 (1953) for oxalyl chloride, and G. I. M. Broom and L. E. Sutton, J. Chem. Soc., 1941, 727, and J. E. LuValle and V. Schomaker, J. Am. Chem. Soc., 61, 3520 (1939) for diacetyl.

11a. J. V. Quagliano and S. Mizushima, J. Am. Chem. Soc., 75 6084 (1953).

can be concluded to be the *gauche*, although the *cis* configuration has so far been tacitly assumed for such kind of ligands in coordination compounds. Ethylene diamine ligand in coordination complexes has also been shown to be in the *gauche* configuration.

The conclusion of the *gauche* configuration is of interest to the stereo-chemistry of these complexes, for the *gauche* configuration has a mirror image non-superposable. Thus more optical isomers are theoretically possible, if the chelating group present in these complexes has the *gauche* rather than the *cis* configuration.[11b]

## 14. Cyclic Compounds

Carbon atoms can form rings connected by single bonds, in which these atoms tend to keep the tetrahedral valency. For such ring compounds we have some interesting problems relating to internal rotation, of which we shall cite a few examples in the following.

Cyclopentane has usually been considered to have a planar configuration, but according to the determination of the entropy by Aston et al.[12] the symmetry number of the molecule cannot be as great as 10 and accordingly the carbon skeleton is not planar. Pitzer[13] pointed out that in the planar configuration the adjacent methylene groups are all oriented in eclipsed positions with respect to internal rotation about the C – C axis. Following this Kilpatrick, Pitzer and Spitzer[14] discussed the nature and the extent of the deviation of the molecule from a planar structure and the effect of this on the thermodynamic functions. We see from these studies that the distortion of the C – C – C bond angle to a certain extent from the tetrahedral value is easier than keeping the eclipsed rotational configuration. This is compatible with the previous conclusion that the eclipsed form of ethane is unstable and that the staggered form corresponds to the stable configuration.

---

11b. The X-ray and the optical measurements have also shown that the ethylene diamine ligand has the *gauche* configuration in coordination complexes. See T. Watanabe and M. Atoji, Kagaku (Science), **21**, 301 (1951); A. Nakahara, Y. Saito and H. Kuroya, Bull. Chem. Soc. Japan, **25**, 331 (1952); M. Kobayashi, J. Chem. Soc. Japan, **64**, 648 (1943).

12. J. G. Aston, S. C. Schumann, H. L. Fink and P. M. Doty, J. Am. Chem. Soc., **63**, 2029 (1941).

13. K. S. Pitzer, Science, **101**, 672 (1945).

14. J. E. Kilpatrick, K. S. Pitzer and R. Spitzer, J. Am. Chem. Soc., **69**, 2483 (1947).

On the basis of the tetrahedral valency of the carbon atom two forms of the cyclohexane molecule would be possible. Four carbon atoms lie in a plane in either form. However, the other two carbon atoms lie one above and one below this plane in the "chair form", and both lie above this plane in the "boat form" (Fig. 4.5).

Chair form                    Boat form

Fig. 4.5. The two configurations of the cyclohexane molecule.

An electron diffraction investigation of this substance was made by Hassel et al.[15] and spectroscopic measurements have been made by many investigators[16] including Mizushima, Morino and Takeda. The experimental results show that substantially all the molecules are in the chair form. The assignment of the Raman and infrared frequencies to the vibrational modes of the chair form has been made by Mizushima, Morino and Fujishiro,[17] by Beckett, Pitzer and Spitzer[18] and by Ramsay and Sutherland.[19] Shimanouchi and Kojima[20] made the normal vibration calculation on the basis of the Urey-Bradley field. In Table 4.1 are shown these calculated frequencies which are in good agreement with those observed in the Raman effect and infrared absorption.

The experimental conclusion stated above is quite reasonable from the internal rotation potential explained in the preceding sections, because in the chair form the internal rotation maintains the staggered position throughout, while the boat form twists two C – C bonds into the eclipsed configuration. From this fact Pitzer calculated the energy difference between the chair and boat forms as twice the internal rotation potential barrier in ethane or 5.6 kcal./mole. Beckett, Pitzer and Spitzer made a thermodynamic

15. O. Hassel, Tids. Kjemi, Bergvesen Met., 3, 32 (1943).
16. See, for example, R. S. Rasmussen, J. Chem. Phys., 11, 249 (1943).
17. S. Mizushima, Y. Morino and R. Fujishiro, J. Chem. Soc. Japan, 62, 587 (1941).
18. C. W. Beckett, K. S. Pitzer and R. Spitzer, J. Am. Chem. Soc., 68, 2488 (1947).
19. D. A. Ramsay and G. B. B. M. Sutherland, Proc. Roy. Soc., (A) 190, 245 (1947).
20. T. Shimanouchi and K. Kozima, J. Chem. Soc. Japan, 72, 468 (1951).

investigation of this substance and concluded that at room temperature cyclohexane is predominantly in the chair form, but that the specific heat of the gas shows a contribution from conversion to the boat form.[18]

TABLE 4.1

Skeletal frequencies of the chair form of cyclohexane in cm.$^{-1}$
($p$: polarized; $d$: depolarized)

| Type of Vibration | | calc. | obs. | |
|---|---|---|---|---|
| $A_{1g}$ | deformation | 384 | 382 | Raman ($p$) |
| | stretching | 805 | 802 | Raman ($p$) |
| $A_{1u}$ | stretching | 1068 | — | |
| $A_{2u}$ | deformation | 674 | 676 | infrared |
| $E_g$ | deformation | 421 | 426 | Raman ($d$) |
| | stretching | 1039 | 1028 | Raman ($d$) |
| $E_u$ | deformation | 208 | — | |
| | stretching | 901 | 907 | infrared |

In the more stable chair form there are two kinds of hydrogen atoms (Fig. 4.6). Three hydrogen atoms are distinctly above the general plane of the carbon atoms and three are distinctly below this plane. These are named by Pitzer as "polar" hydrogen atoms*. The other six form a circle outside the carbon atoms and in the same general plane as the carbon atoms. These are called by him "equatorial" hydrogen atoms. Therefore, of the two hydrogen atoms on each carbon, one is polar and the other is equatorial. If the six carbon atoms pass through a single plane over to the opposite chair configuration all hydrogens originally equatorial become polar and vice versa. So long as all are hydrogen atoms, nothing new is obtained

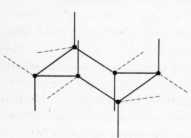

Fig. 4.6. The chair form of cyclohexane. (Full lines denote polar bonds and dotted lines equatorial bonds.)

by this process but with other groups attached, different molecular configurations become possible. For example, in methylcyclohexane the methyl

* Recently it has been suggested to use "axial" in place of "polar" by D.H.R. Barton, O. Hassel, V. Prelog and K. S. Pitzer. (Private communication of Professor Prelog.)

group may be in a polar position or in an equatorial position. From the similarity of the polar configuration to the *gauche* configuration of *n*-butane, Pitzer et al. estimated the energy of the polar form to be 1.8 kcal./mole higher than that of the equatorial form with on *gauche* configuration at all. Accordingly, at room temperature most methylcyclohexane molecules will take a chair form with an equatorial methyl group. Similar experimental results have been obtained for monochlorocyclohexane by Hassel[21] who showed by the electron diffraction method the predominance of molecules with the chlorine atom in an equatorial position.

For dichlorocyclohexane Kozima and Yoshino made the Raman and dipole measurements.[22] For 1,2-disubstituted compounds four combinations of C–Cl bonds are conceivable, i. e. $(1\,p, 2\,p)$, $(1\,e, 2\,e)$, $(1\,p, 2\,e)$ and $(1\,e, 2\,p)$ where $p$ denotes the polar bond and $e$ the equatorial bond. By the distortion of the ring described above, $p$ is changed to $e$ and vice versa. Accordingly, we have only two different structure which may be expressed as

$$(1\,p, 2\,p) \rightleftarrows (1\,e, 2\,e)$$
$$(1\,p, 2\,e) \equiv (1\,e, 2\,p)$$

According to the experimental results of Kozima et al., *trans*-1,2-dichloro-cyclohexane in solutions exists in the former configurations, of which only $(1\,e, 2\,e)$ persists in the solid state. Similar results were obtained for *trans*-1,4-dichlorocyclohexane.[22] In the solutions the equilibrium

$$(1\,p, 4\,p) \rightleftharpoons (1\,e, 4\,e)$$

was observed and in the solid state only the configuration $(1\,e, 4\,e)$ was found to persist. Corresponding electron diffraction and X-ray measurements were made by Hassel et al. who have studied extensively the molecular configuration of cyclohexane derivatives. (See Table 4.2.)

In connection with the stable molecular forms of cyclohexane, we should like to comment on those of 1,4-dioxane which may exist in various forms shown in Fig. 4.7. According to the dipole measurement by Kubo[23] in the gaseous state and by Williams[24] and others[25] in solutions, the moment value is 0.4 D. The small but finite moment would indicate that the greater part of the molecules exist in the nonpolar chair form, yet nevertheless the number

21.  O. Hassel, Research **3**, 504 (1950).
22.  K. Kozima and T. Yoshino, J. Chem. Soc. Japan, **72**, 20 (1951).
23.  M. Kubo, Sci. Pap. Inst. Phys. Chem. Res. Tokyo, **30**, 238 (1936).
24.  J. W. Williams, J. Am. Chem. Soc., **52**, 1831 (1930).
25.  R. Sängewald and A. Weissberger, Phys. Z., **30**, 268 (1929); E. C. E. Hunter and J. R. Partington, J. Chem. Soc., **1933**, 87.

of molecules in polar boat forms is not negligibly small. However, Yasumi and Shirai[26] have concluded from their measurement by very short radio waves that the moment value is much less than this. Furthermore, the

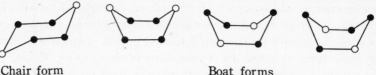

Chair form                    Boat forms

Fig. 4.7. Possible stable configurations of 1,4-dioxane.

experimental results of the Raman and infrared measurements have also shown that those molecules in boat form are present in an almost negligible amount.*

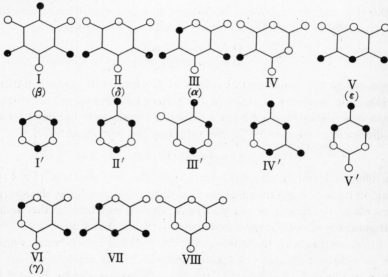

Fig. 4.8. Molecular configurations of the isomers of benzene hexachloride.

The moment value $(0.4\ D)$ referred to above may, therefore, arise from the atomic polarization, and the molecule of 1,4-dioxane may have no moment at ordinary temperature. If such is actually the case, we have to

---

* See for example, F. E. Malherbe and H. J. Bernstein, J. Am. Chem. Soc., 74, 4408 (1952). See also K. W. F. Kohlrausch and W. Stockmair, Z. phys. Chem. (B) 31, 382 (1935).

---

26. M. Yasumi and M. Shirai, Private Communication.

conclude that almost all the molecules are in the chair form. We shall come back to this problem in Part II, after we have explained the meaning of the atomic polarization in connection with the absorption frequencies in the infrared region.

TABLE 4.2

Molecular configurations of some cyclohexane derivatives
($e$: equatorial, $p$: polar)

| Substance | State | Configuration | |
|---|---|---|---|
| monomethycyclohexane | vapor | $1e$ | Pitzer et al. |
| monochlorocyclohexane | vapor | $1e$ | Hassel et al. |
| *trans*-1,2-dichlorocyclo-hexane | solid | $1e, 2e$ | Kozima et al. |
| | nonpolar solution | $1p, 2p \rightleftarrows 1e, 2e$ | Kozima et al. |
| *trans*-1,4-dichlorocyclo-hexane | solid | $1e, 4e$ | Kozima et al. |
| | nonpolar solution | $1p, 4p \rightleftarrows 1e, 4e$ | Kozima et al. |
| | vapor | $1e, 4e$ | Hassel et al. |
| *trans*-1,3-dibromocyclo-hexane | vapor | $1e, 3e$ | Hassel et al. |
| $\alpha$-hexachlorocyclohexane | solid and vapor | $1p, 2e\ 3e\ 4e\ 5e\ 6p$ | Hassel et al. |
| | nonpolar solution | $1p, 2e\ 3e\ 4e\ 5e\ 6p$ | Morino et al. |
| $\beta$-hexachlorocyclohexane | solid and vapor | $1e, 2e\ 3e, 4e, 5e, 6e$ | Hassel et al. |
| | nonpolar solution | $1e, 2e, 3e, 4e, 5e, 6e$ | Morino et al. |
| $\gamma$-hexachlorocyclohexane | vapor | $1p, 2e, 3e, 4e, 5p, 6p$ | Hassel et al. |
| | nonpolar solution | $1p, 2e, 3e, 4e, 5p, 6p$ | Morino et al. |
| $\delta$-hexachlorocyclohexane | solid and vapor | $1p, 2e, 3e, 4e, 5e, 6e$ | Hassel et al. |
| | nonpolar solution | $1p, 2e, 3e, 4e, 5e, 6e$ | Morino et al. |
| $\varepsilon$-hexachlorocyclohexane | solid and vapor | $1p, 2e, 3e, 4p, 5e, 6e$ | Hassel et al. |
| | nonpolar solution | $1p, 2e, 3e, 4p, 5e, 6e$ | Morino et al. |

Benzene hexachloride or 1,2,3,4,5,6-hexachlorocyclohexane may exist in thirteen configurations shown in Fig. 4.8, even if the carbon ring takes only the chair form. Of these thirteen configurations we can find the following five pairs, one configuration of each pair being convertible to another by the distortion process described above.

$$1e, 2e, 3e, 4e, 5e, 6e \quad (\text{I}) \quad \rightleftharpoons \quad 1p, 2p, 3p, 4p, 5p, 6p \quad (\text{I}')$$
$$1p, 2e, 3e, 4e, 5e, 6e \quad (\text{II}) \quad \rightleftharpoons \quad 1e, 2p, 3p, 4p, 5p, 6p \quad (\text{II}')$$
$$1p, 2e, 3e, 4e, 5e, 6p \quad (\text{III}) \rightleftharpoons \quad 1e, 2p, 3p, 4p, 5p, 6e \quad (\text{III}')$$
$$1p, 2e, 3p, 4e, 5e, 6e \quad (\text{IV}) \rightleftharpoons \quad 1e, 2p, 3e, 4p, 5p, 6p \quad (\text{IV}')$$
$$1p, 2e, 3e, 4p, 5e, 6e \quad (\text{V}) \quad \rightleftharpoons \quad 1e, 2p, 3p, 4e, 5p, 6p \quad (\text{V}')$$

The remaining three configurations are

$$1\,p,\,2\,e,\,3\,e,\,4\,e,\,5\,p,\,6\,p \quad \text{(VI)}$$
$$1\,p,\,2\,e,\,3\,e,\,4\,p,\,5\,e,\,6\,p \quad \text{(VII)}$$
$$1\,p,\,2\,e,\,3\,p,\,4\,e,\,5\,p,\,6\,e \quad \text{(VIII)}$$

The distance between two chlorine atoms, one in one $p$-bond and the other in every second $p$-bond is 2.52Å which is far less than the van der Waals distance between two chlorine atoms (3.60Å). The latter five configurations of the five pairs contain more such pairs of chlorine atoms than the former five configurations, and the latter ones are considered less stable. This means that we may consider eight stable configurations in all. Actually all the configurations so far determined by electron diffraction experiments by Hassel et al.,[21] by dipole measurements by Hassel et al.[27] and Morino, Miyagawa and Oiwa[28] and by infrared and Raman measurements by Daasch[29] and Kuratani, Sakashita, Takeda and Mizushima[30, 31] correspond to five of these eight configurations. (See Table 4.2.) In calculating the dipole moment of these eight configurations Morino, Miyagawa and Oiwa[28] used a new empirical method by which they had been able to calculate moment values of simple chlorinated paraffins in excellent agreement with the observed values. In Table 4.3 are shown the computed and the observed moment values of benzene hexachloride.

TABLE 4.3

Dipole moment of benzene hexachloride

| Isomer | Calculated | Observed |
|--------|------------|----------|
| $\alpha$ | 2.25 D | 2.22 D |
| $\beta$ | 0 | 0 |
| $\gamma$ | 2.93 | 2.80 |
| $\delta$ | 2.25 | 2.22 |
| $\varepsilon$ | 0 | 0 |

27. O. Hassel and E. Naeshagen, Z. phys. Chem., **B 14**, 79 (1931).

28. Y. Morino, I. Miyagawa and T. Oiwa, Botyu-Kagaku, **15**, 181 (1950).

29. L. W. Daasch, Anal. Chem., **19**, 779 (1947).

30. K. Kuratani, T. Shimanouchi and S. Mizushima, Rep. Rad. Chem. Res. Inst. Tokyo Univ., **3**, 16 (1948).

31. K. Kuratani, K. Sakashita and S. Takeda, ibid., **5**, 8 (1950).

## 15. Internal Rotation about the C-O Axis

The simplest molecule containing a C–O bond as internal rotation axis is methyl alcohol (Fig. 4.9), the Raman and infrared spectra of which have been measured by a number of investigators. The Raman band of the liquid observed at 3270–3480 cm.$^{-1}$ is assigned with certainty to the OH stretching frequency, but there was once ambiguity as to the frequency value of the corresponding bending vibration. Mizushima, Morino and Okamoto[32] were able to show by the measurement of the isotopic shift that the 1107 cm.$^{-1}$ line corresponds to this bending frequency.* (This is shifted to 955 cm.$^{-1}$ in $CH_3OD$.) The third OH frequency which corresponds to the torsional motion of the OH group about the C – O axis is much lower than this bending frequency. Evidently this is most closely related to the internal rotation potential which has a threefold symmetry just as in the case of ethane.

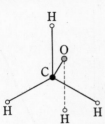

Fig. 4.9. The stable configuration of methyl alcohol.

In 1926–1928 Mizushima[33, 34, 35] published a series of papers on the dispersion of electric waves by monovalent alcohols in the wave-length range from 50 m to 50 cm. It was found that in a certain range of wave length the dielectric constant drops down from the large value equal to the static dielectric constant to the small value almost equal to the square of refractive index for visible light (see Fig. 4.10). At that time the experimental results were explained by Debye's theory with the picture of dielectric liquids as consisting of spherical molecules whose orientation in an externally applied field becomes less complete at higher frequency.[36] Later on Mizushima and Kubo[37] continued this kind of measurement in highly viscous media in order

---

* In the gaseous state we have much less intermolecular hydrogen bonding and, therefore, the bending vibration appears at higher frequency, 1340 cm.$^{-1}$ according to H. D. Noether, J. Chem. Phys., **10**, 693 (1942). Undoubtedly there is a coupling of this vibration with C – O stretching and C – H bending vibrations. [See K. Kuratani, J. Chem. Soc. Japan, **73**, 758 (1952).]

32. S. Mizushima, Y. Morino and G. Okamoto, Bull. Chem. Soc. Japan, **11**, 698 (1936).
33. S. Mizushima, Bull Chem. Soc. Japan, **1**, 47, 83, 115, 143, 163 (1926).
34. S. Mizushima, Physik. Z., **28**, 418 (1927).
35. S. Mizushima, Sci. Pap. Inst. Phys. Chem. Res. Tokyo, **5**, 201 (1927), **9**, 209 (1928).
36. P. Debye, Verh. Deut. Phys. Ges., **15**, 770 (1913); Polar Molecules, Chemical Catalogue Company, New York (1929).
37. S. Mizushima and M. Kubo, J. Chem. Soc. Japan, **62**, 502 (1941).

to estimate the height of the potential barrier to internal rotation.  If the overall rotation of the molecule is small as compared with the internal rotation, we may estimate the barrier height from the dispersion phenomena by use of the absolute reaction rate theory.   In this way Mizushima and Kubo obtained a value which is at least correct in the order of magnitude.[37]

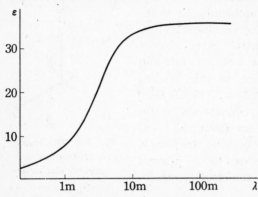

Fig. 4.10.  The wave length dependence of the dielectric constant $\varepsilon$ of ethyl alcohol at –40° C.

More accurate determination of the barrier height by microwave spectroscopy can be made in the shorter wave length region.   In methyl alcohol, the torsional motion is spectroscopically active and, transitions between components of torsional levels can be observed with overall rotational changes.   These components are considered to arise from the quantum mechanical tunneling phenomena associated with the existence of three equivalent potential minima. Based on this consideration Burkhard and Dennison[38] estimated the barrier height as about 1000 cal./mole, assuming a sinusoidal potential function, as in the case of ethane.

The potential barrier to internal rotation of methyl alcohol can also be obtained from the difference between the third law entropy and molecular entropy for free rotation.   Such a method has been applied by Pitzer,[39] Crawford,[40] Halford,[41] and others.   The value obtained by Halford[41] is 1600 ± 700 cal./mole. While the range of the thermodynamic value includes the microwave value, it seems to be too high.   Any undetermined entropy in the crystal makes the barrier appear higher.   Halford attributed the difference to possible zero point entropy at the absolute zero due to hydrogen bonding.  (Hydroxyl compounds such as methyl alcohol may contain residual entropy at the absolute zero due to the random orientation of hydroxyl

38.  D. G. Burkhard and D. M. Dennison, Ohio State Symposia on Molecular Structure (1948 and 1949); Phys. Rev., 84, 408 (1951).

39.  K. S. Pitzer, J. Am. Chem. Soc., 70, 2140 (1948).

40.  B. L. Crawford, J. Chem. Phys., 8, 744 (1940).

41.  J. O. Halford, J. Chem. Phys., 18, 361 (1950).

ponds.) However, he pointed out also that uncertainty in the gas-imperfection correction may be responsible for the difference between the two values of barrier height.

This problem was also discussed by Weltner and Pitzer[42] who found by the measurement of heat capacity of methyl alcohol vapor that there is unexpected polymerization of methyl alcohol vapor. By taking proper account of this gas-phase polymerization they could bring the experimental thermodynamic quantities into agreement with values calculated from the microwave value of about 1000 cal./mole. However, they consider that due to the uncertainty in the assignment of vibrational frequencies, the thermo-dynamic data might be consistent with any barrier height within a few hundred calories of this 1000 cal. figure.

Similarly the potential barrier of ethyl alcohol has been discussed from the experimental thermodynamic quantities.[43] According to the recent work of Barrow,[44] barriers hindering internal rotation of $CH_3$ and OH groups of about 3000 and 1000 cal., respectively, are in agreement with the observed heat capacities, entropies and constants for the equilibrium of water, ethylene and ethyl alcohol. In this case the correction of gas imperfection is the most important factor in treating the observed thermodynamic data, as in the case of methyl alcohol. Thus it seems that the most probable value of the barrier hindering the internal rotation of OH group is about 1000 cal./mole.

The OH frequencies referred to in the beginning of this section refer to the liquid state where we have to expect strong *intermolecular* hydrogen bonds. Accordingly, we have to use the frequencies of free molecule in the calculation of entropy except at low temperatures, at which the difference between the OH frequencies in the free and hydrogen bonded molecules is insignificant in this calculation. However, in the case of *intramolecular* hydrogen bonds, as in the case of the *gauche* molecule of ethylene chlorhydrin stated in Section 11, even the free molecule exhibits the hydrogen-bonded OH frequencies. We have seen that due to this *intramolecular* hydrogen bond in ethylene chlorhydrin one of the potential minima in the internal rotation about the C–O axis becomes much more stable than the other two, but this does not at all mean the latter two potential minima can be neglected. For example, if we determine the energy difference between the *trans* and

---

42. W. Weltner and K. S. Pitzer, J. Am. Chem. Soc., **73**, 2606 (1951).

43. Most of the earlier work has been discussed by J. O. Halford, J. Chem. Phys., **17**, 111 (1949), **18**, 361 (1950). See also K. Ito, J. Chem. Phys., **20**, 531 (1952).

44. G. M. Barrow, J. Chem. Phys., **20**, 1739 (1952).

the *gauche* molecules of ethylene chlorhydrin from the intensity measurement of the free and hydrogen-bonded OH bands, this would lead to an erroneous result. This measurement yields an energy difference of 2.0 ± 0.5 kcal./mole (Zumwalt and Badger)[45] which is larger than the value (0.95 kcal./mole obtained by Mizushima et al.[46] (See Section 11.) This difference can be explained as follows.

As the energy difference shown in Section 11 has been obtained from the intensity measurement of the absorption bands arising from the skeletal motions, this refers to the difference between the weighted mean energy of the *trans* molecule and that of the *gauche* molecule, because the skeletal frequencies are not affected appreciably by the difference in the position of the hydrogen atom. However, the energy difference obtained from the absorption intensity of the OH band is the difference between the weighted mean energy of the *trans* and the *gauche* molecules without hydrogen bond and that of the *gauche* molecule with internal hydrogen bond. Therefore, it is no wonder that we have two different values of energy difference, so long as the number of molecules with OH group in the less stable positions of the internal rotation about the C – O axis is not very small as compared with that in the most stable position.

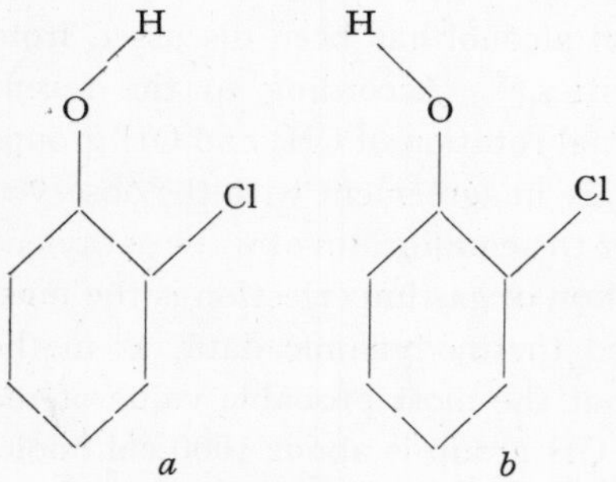

Fig. 4.11.
Two configurations of o-chlorophenol.

The internal rotation about the C – O axis in other alcohols should not be much different from that in methyl alcohol, but when the OH group is directly coupled with a benzene nucleus, the situation becomes quite different, because in this case the C – O bond aquires the double bond character due to resonance. The simplest molecule of this kind is phenol $C_6H_5OH$, for which we assume two identical planar configurations instead of the free rotation of OH group about the C–O axis. It would not be simple to prove experimentally that such is actually the case for phenol, but if we replace another hydrogen atom of phenol by a suitable atom or group, the

45. L. R. Zumwalt and R. M. Badger, J. Am. Chem. Soc., 62, 305 (1940).

46. S. Mizushima, T. Shimanouchi, T. Miyazawa, K. Abe and M. Yasumi, J. Chem. Phys., 19, 1477 (1951).

experimental determination of the existence of rotational isomers may become quite simple. For example, in orthochlorophenol we can consider the two configurations shown in Fig. 4.11 due to the partial double bond character of the C–O bond which tends to keep the hydroxyl hydrogen on the same plane as the benzene nucleus. In form $b$ the hydroxyl hydrogen does not interfere with the chlorine atom, while it does in form $a$, so that the OH frequency of form $a$ is shifted towards lower frequency as first suggested by Pauling.[47] This was clearly shown by the infrared measurement of the first overtone by Wulf and Liddel[48] and of the second and third overtones by Mizushima, Uehara and Morino.[49]   Besides this substance several other compounds with hydroxyl groups have been measured by these investigators and some of the experimental results are shown in Table 4.4.

TABLE 4.4
Frequencies of OH absorption peaks in cm.$^{-1}$ (Solvent: $CCl_4$)

| Substance | $\omega$ | $x\,\omega$ | $\nu_1$ calc. | $\nu_1$ obs. | $\nu_2$ | $\nu_3$ |
|---|---|---|---|---|---|---|
| $C_6H_5OH$ | 3790 | 87 | 7061 | 7050 | 10332 | 13430 |
| $p$-$C_6H_4ClOH$ | 3773 | 83 | 7047 | | 10322 | 13430 |
| $o$-$C_6H_4ClOH$ | 3701 | 84 | 6897 | 6910 | 10094 | 13122 |
| | 3775 | 84 | 7049 | 7050 | 10323 | 13430 |
| $o$-$C_6H_4BrOH$ | 3705 | 93 | 6855 | 6860 | 10005 | 12970 |
| | 3812 | 92 | 7070 | 7050 | 10328 | 13401 |
| $o$-$C_6H_4CH_3OH$ | 3783 | 85 | 7055 | 7060 | 10328 | 13430 |

It is quite reasonable from the point of view stated above that only one absorption peak is observed for such a molecule as $o$–$C_6H_4CH_3OH$ or $p$-$C_6H_4ClOH$.

If we express approximately the vibrational level of the quantum number $v$ as:

$$G(v) = \omega\left(v + \frac{1}{2}\right) - x\,\omega\left(v + \frac{1}{2}\right)^2,$$

we can calculate the frequency of the infinitesimal vibration $\omega$ and the anharmonicity factor $x\,\omega$ from the observed frequencies of the second and third overtones as shown in Table 4.4.[49] From the values of $\omega$ and $x\,\omega$ we

47. L. Pauling, J. Am. Chem. Soc., **58**, 94 (1936).
48. O. R. Wulf and U. Liddel, J. Am. Chem. Soc., **57**, 1464 (1935).
49. S. Mizushima, Y. Uehara and Y. Morino, Bull. Chem. Soc. Japan, **12**, 132 (1937).

can also calculate the frequencies of the first overtone, which are in good agreement with the observed values. Mizushima, Kubota and Morino made a similar measurement in the gaseous state with essentially the same result.[50] (There is a small shift of absorption peak in the gaseous state.)

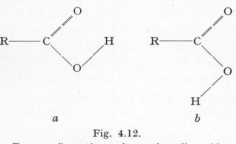

a                    b
Fig. 4.12.
Two configurations of a carboxylic acid.

The dipole moment of o-chlorophenol has been measured by Zahn,[51] Williams,[52] Kozima and Mizushima[53] and Curran et al.[54] The observed value is close to that calculated for form a of Fig. 4.11 and further, there was found no temperature-dependence of the moment over a wide range of temperature. From this result it will be seen that the molecule of o-chlorophenol is predominantly in form a.

Similar structures are expected in the case of carboxylic acids. Due to single bond-double bond resonance we can consider the two configurations shown in Fig. 4.12. In contradistinction to the case of o-chlorophenol, the infrared measurements show no absorption peak to be assigned to configuration b of Fig. 4.12 (Mizushima, Kubota and Morino).[50] The dipole measurement made by Zahn[55] in the gaseous state is compatible with this result. The difference between the calorimetric entropy and the molecular entropy not including the contribution of the torsional vibration of the OH group was found to be 0.791 e. u.[56] From this the frequency of the torsional motion is calculated as 452 cm.$^{-1}$. This corresponds to a barrier of about 10 kcal./mole.

As is well known, some carboxylic acids exist in double and single molecules even in the gaseous state. For example Pauling and Brockway[57] showed by electron diffraction measurements that formic acid is associated

50. S. Mizushima, T. Kubota and Y. Morino, Bull. Chem. Soc. Japan, 14, 15 (1939).
51. C. T. Zahn, Phys. Rev., 27, 455 (1926).
52. J. W. Williams, Physik. Z., 29, 683 (1928); J. Am. Chem. Soc., 52, 1356 (1930).
53. K. Kozima and S. Mizushima, Sci. Pap. Inst. Phys. Chem. Res. Tokyo, 31, 296 (1937).
54. W. F. Anzilotti and B. Columba Curran, J. Am. Chem. Soc., 65, 607 (1943).
55. C. T. Zahn, Trans. Farad. Soc., 30, 804 (1934).
56. W. Warning, Chem. Rev., 51, 171 (1952); J. O. Halford, J. Chem. Phys., 10, 582 (1942); 14, 395 (1946). J. W. Stout and L. H. Fisher, J. Chem. Phys., 9, 163 (1941).
57. L. Pauling and L. O. Brockway, Proc. Nat. Acad. Sci., 20, 336 (1934); J. Karle and L. O. Brockway, J. Am. Chem. Soc., 66, 574 (1944).

through the hydrogen bonding to give a structure $H-C\begin{smallmatrix}O-H\;O\\O\cdots H-O\end{smallmatrix}C-H$.

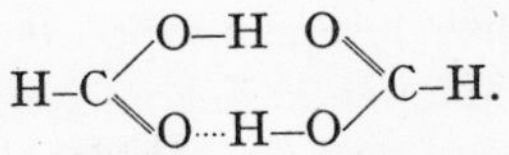

In the non-polar solutions formic acid and other carboxylic acids are also considered to be associated as dimers. However, in the solid state these acid molecules seem to be associated in a different manner. The X-ray diffraction investigation by Holtzberg, Post and Fankuchen[58] seems to prelude the possibility that the molecules are associated as dimers in the crystal. Preliminary calculations indicate that the molecules are arranged in the crystal in the form of infinite chains, each molecule being linked to two neighbors by hydrogen bonds.

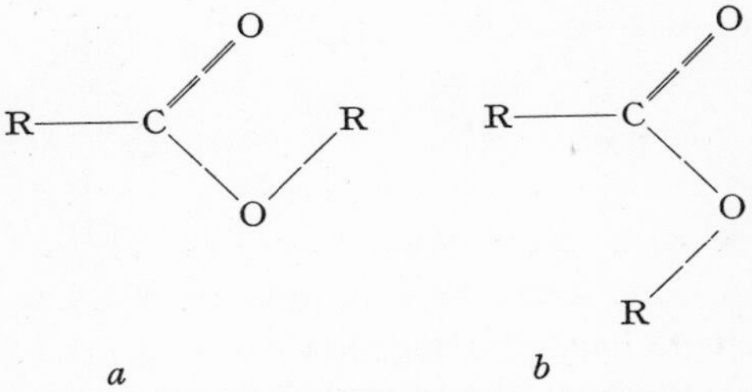

There may be some inhibiting effect on the carboxylic resonance if the hydrogen atom of the acid molecule is replaced by an alkyl radical to form the ester molecule, for which we may again consider the two configurations shown in Fig. 4.13. The dipole measurements on these esters have been made by several authors, including Mizushima and Kubo who found for the vapor of methyl acetate a constant value 1.70 D over a temperature range from 35° to 210° C.[59] This value together with its temperature-independence shows that the resonance effect is in this case still considerable and the ester molecules are locked at the position of $a$ of Fig. 4.13. This

Fig. 4.13.

Two configurations of a carboxylic ester.

TABLE 4.5

Dipole moment of methyl and ethyl chloroformate[59]

| Methyl chloroformate | | Ethyl chloroformate | |
|---|---|---|---|
| $t$ °C | $m$ (D) | $t$ °C | $m$ (D) |
| 34.6 | 2.38 | 35.0 | 2.56 |
| 78.0 | 2.40 | 76.9 | 2.47 |
| 140.2 | 2.29 | 137.6 | 1.79 |
| 207.4 | 1.68 | 207.3 | 1.43 |

58. F. Holtzberg, B. Post and I. Fankuchen, J. Chem. Phys., 20, 198 (1952).
59. S. Mizushima and M. Kubo, Bull. Chem. Soc. Japan, 13, 174 (1938).

conclusion is supported by the electron diffraction experiment by O'Gorman, Shand and Schomaker.[60] If, however, we replace one methyl group of methyl acetate by a chlorine atom to form methyl chloroformate the dipole moment is found to change considerably with temperature according to the measurement of Mizushima and Kubo[59]. (See Table 4.5.) This is also the case for ethyl chloroformate. It is worthy of note that there was found no appreciable change of Raman spectrum of this substance between the liquid and solid states[61] and it seems that the spectrum can be explained by assuming only one molecular configuration, which according to the electron diffraction experiment referred to above [60] is approximately the *cis* form (form *a* of Fig. 4.14). In order to explain the temperature-dependence of dipole moment, it seems at first sight necessary to consider a rather large amplitude of oscillation about the equilibrium azimuthal angle.[60] However, it is possible that the dipole moment of the molecule changes considerably with azimuthal angle $\theta$ due to the change of the resonance with $\theta$, so that the oscillation of not very large amplitude may account for the observed temperature-dependence of dipole moment. This view is compatible with the result of the Raman measurement.

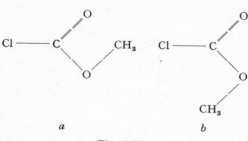

*a*                     *b*

Fig. 4.14.
Configurations of methyl chloroformate.

In Table 4.6 are shown the results of dipole measurement on isomers of dimethoxybenzene.[62] If the two $OCH_3$-groups of quinol dimethyl ether (*para*-compound) rotate freely about the $O-C_6H_4-O$ axis, the dipole moment will be independent of temperature and its value is calculated as

$$m = \sqrt{2}\, \mu \sin \alpha \qquad (4.1)$$

where $\mu$ is the C–O bond moment and $\alpha$ the oxygen valence angle. Putting $\mu = 1.1$ D and $\alpha = 110°$, we have

$$m = 1.46 \, \text{D}.$$

60. J. M. O'Gorman, W. Shand and V. Schomaker, J. Am. Chem. Soc., **72**, 4222 (1950).

61. K. Kurosaki, I. Ichīshima and S. Mizushima, unpublished.

62. S. Mizushima, Y. Morino and H. Okazaki, Sci. Pap. Inst. Phys. Chem. Res. Tokyo, **34**, 1147 (1938).

TABLE 4.6

Dipole moment of $C_6H_4(OCH_3)_2$ measured by Mizushima, Morino and Okazaki in benzene[62]

| $t \, °C$ | ortho | meta | para |
|---|---|---|---|
| 40 | 1.32 $D$ | 1.59 $D$ | 1.71 $D$ |
| 25 | 1.24 $D$ | 1.58 $D$ | 1.70 $D$ |
| 10 | 1.18 $D$ | 1.57 $D$ | 1.69 $D$ |

The difference between this and the observed value (see Table 4.6) may be attributable to the approximation in calculating the resultant moment. From this point of view the temperature-dependence of the moment value of veratrol (*ortho*-compound) is explained by the steric and electrostatic interaction between the two $OCH_3$-groups.

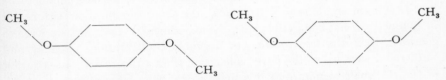

Fig. 4.15. Configurations of quinol dimethyl ether.

If, however, the C–O bond partially acquires double bond character, the methyl groups tend to be fixed in the plane of the benzene nucleus to yield the two configurations shown in Fig. 4.15. The dipole moment of the former is zero, while that of the latter is $2\mu \sin \alpha$. Hence, if these two forms have the same energy, they will exist in the same amount and the mean moment is calculated as

$$m = \sqrt{\frac{1}{2} \times 4 \, \mu^2 \sin^2 \alpha} = \sqrt{2} \, \mu \sin \alpha$$

This is identical with the expression (4.1) for the free rotation, so that the second view will explain the experimental result as well as the first.

We might take the third view in which we consider two stable configurations in the absence of appreciable double bond character of the C–O bond. The hydrogen atoms attached to the benzene nucleus may exert a considerable resistance against the internal rotation of $OCH_3$ groups about the $O–C_6H_4–O$ axis, so that all the methyl groups of quinol dimethyl ether are situated at the farthest distance apart from the benzene plane. The dipole moment values of these two forms are exactly the same as in the second case and accordingly, if we assume the same energy for these two forms, we can explain the experimental result equally well.

Mizushima, Morino and Okazaki have also made Raman measurements on quinol dimethyl ether in the liquid and solid states.[62] They found a difference between the spectra of these two states from which we may consider two configurations in the liquid state and only one in the solid state. However, no theoretical treatment of the spectra has yet been made to determine conclusively which of the views proposed from the dipole measurement is most reasonable from the spectroscopic point of view.

### 16. Internal Rotation about O-O, S-S and Si-Si bonds as axes

The simplest molecule with an O–O bond as internal rotation axis is hydrogen peroxide HO–OH, the dipole moment of which was found to be 2.13 $D$ (Linton and Maass).[63] Therefore, the *trans* configuration with center of symmetry is not compatible with the experimental result and we have to consider either the free rotation or a *gauche* configuration. Penney and Sutherland were led to the latter conclusion from a theoretical investigation of the electronic structure of this molecule.[64] The essential feature of the argument is that the electrons of the two oxygen atoms arrange themselves to form the strongest possible bonds and in so doing render the charge density on these atoms unsymmetrical about the O–O axis. The interaction of these two electronic clouds is the dominant factor in determining the azimuth of one group relative to the other which is approximately 90° (Fig. 4.16): all other types of interaction (e. g. dipole interaction) have been estimated to shift the equilibrium position only slightly.

The infrared absorption has been observed by many investigators, including the recent measurement of Giguère who proposed new assignments in agreement with the model referred to above.[65] The same author assigned a band observed at 550 cm.$^{-1}$ to the torsional motions of the OH groups about the O–O axis and estimated the barrier height to be of the order of 4 kcal./mole. The doublet character, especially that of the third harmonic

---

63. E. P. Linton and O. Maass, Can. J. Res. **7**, 81 (1932).

64. W. G. Penney and G. B. B. M. Sutherland, Trans. Farad. Soc., **30**, 898 (1934); J. Chem. Phys., **2**, 492 (1934).

65. P. A. Giguère, J. Chem. Phys., **18**, 88 (1950).

O–H frequency is the most remarkable feature of the infrared absorption band of hydrogen peroxide. Evidently this is due to the presence of the double minimum potential associated with the internal rotation of OH groups.

From the similarity of the electronic structure, the molecule of sulphur monochloride Cl–S–S–Cl may be expected to have a *gauche* configuration just as in the case of hydrogen peroxide (Fig. 4.16). Morino and Mizushima have made the Raman and dipole measurements to see if the experimental results can be explained on the basis of this model.[66] The observed moment value of 1.0 D is in good agreement with this configuration and the observed Raman frequencies 104, 208, 242, 445 and 536 cm.$^{-1}$ can be brought into agreement with the calculated values for a molecular model in which the azimuth of one-half of the molecule with respect to the other is about 90°. The result of the electron diffraction experiment by Palmer[67] is in good agreement with the conclusion stated above.

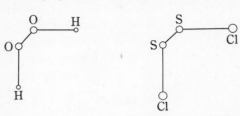

Fig. 4.16. Configurations of hydrogen peroxide and sulphur monochloride viewed along the O–O and S–S axes.

Recently Bernstein and Powling[68] measured the Raman frequencies of this substance with results in good agreement with that obtained by Morino and Mizushima. They also measured the infrared absorption and found the peaks at 438, 448, 538, 779, 881, 984, 1073 and 1333 cm.$^{-1}$. If the five observed frequencies (one infrared and four Raman frequencies):

$$208, \ 242, \ 438, \ 446 \ \text{and} \ 536 \ \text{cm.}^{-1}$$

can be assigned to the fundamental vibrations, the Raman line observed at 104 cm.$^{-1}$ may be due to the torsional motion about the S–S axis.

As stated above the hindering of internal rotation about the O–O or the S–S axis arises from the interaction of the electronic clouds. However, for the Si–Si axis, we have a situation similar to that in the case of C–C axis, as may be seen from the similarity of the valency between carbon and silicon. Indeed there is no reason to doubt that the structure of disilane is

66. Y. Morino and S. Mizushima, Sci. Pap. Inst. Phys. Chem. Res. Tokyo, **32**, 220 (1937).

67. K. J. Palmer, J. Am. Chem. Soc., **60**, 2360 (1938).

68. H. J. Bernstein and J. Powling, J. Chem. Phys., **18**, 1018 (1950).

closely parallel to that of ethane. However, as disilane itself has some experimental disadvantage for structural study just as is the case for ethane, let us discuss in the following the molecular structure of hexachlorodisilane $Cl_3Si–SiCl_3$ in order to obtain information concerning the internal rotation about the Si–Si bond (Fig. 4.17). If the interaction between the two $SiCl_3$ groups is considerable we may expect a stable form with $D_{3d}$ symmetry or a staggered form just as in the case of hexachloroethane $Cl_3C–CCl_3$. If, however, the interaction is very small, the internal rotation will become almost free $(D_{3h}{}')$.

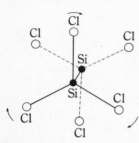

Fig. 4.17.
Hexachlorodisilane
(almost free rotation).

The molecule of hexachlorodisilane has twelve fundamental frequencies, of which six correspond to doubly degenerate vibrations. One of these twelve frequencies is assigned to the internal rotation about the Si–Si bond as axis and is inactive in both the Raman effect and infrared absorption. The selection rule for the remaining eleven frequencies is shown in Table 4.7, in which, besides the rule for the $D_{3d}$ and $D_{3h}{}'$ symmetry, that for the $D_{3h}$ (eclipsed form) is also shown for the sake of comparison. In the same table are shown the calculated frequencies which have been obtained on the basis of the Urey-Bradley field by Katayama, Shimanouchi, Morino and Mizushima[69] and which are in good agreement with the Raman frequencies observed by these investigators, if they are assigned as in the last column of the table.

The most important result of this assignment is that it is not compatible with the $D_{3d}$ symmetry: otherwise the two Raman lines of fairly strong intensity at 588 cm.$^{-1}$ and 179 cm.$^{-1}$ would not appear in the Raman effect, since these two lines correspond to $E_u$ type of $D_{3d}$ and should be inactive. Therefore, so far as these two Raman lines are concerned, the molecule should have the symmetry of $D_{3h}{}'$ or $D_{3h}$ but the latter possibility is ruled out by the electron diffraction investigation by Yamasaki, Kotera, Tatematsu and Iwasaki who concluded that the freely rotating model is in agreement with their experimental result.[70]

69. M. Katayama, T. Shimanouchi, Y. Morino and S. Mizushima, J. Chem. Phys., 18, 506 (1950). The experimental result is in good agreement with that obtained by F. Stitt and D. M. Yost, J. Chem. Phys. 5, 90 (1937).

70. K. Yamasaki, A. Kotera, A. Tatematsu and M. Iwasaki, J. Chem. Soc. Japan, 69, 104 (1947).

TABLE 4.7

Normal vibrations of $Cl_3Si-SiCl_3$

| Vibrational type | | | | Raman activity | | | Vibrational frequencies (cm⁻¹) | | |
|---|---|---|---|---|---|---|---|---|---|
| $D_{3d}$ | $D_{3h}'$ | $D_{3h}$ | | $D_{3d}$ | $D_{3h}'$ | $D_{3h}$ | calc. | obs. | |
| $A_{1g}$ | $A_1$ | $A_1'$ | $\nu$ (Si–Si) | $p$ | $p$ | $p$ | 626 | 622(6) | $p < 0.5$ |
| | | | $\nu$ (Si–Cl) | $p$ | $p$ | $p$ | 375 | 354(10) | $p$ |
| | | | $\nu$ (Si–Cl) | $p$ | $p$ | $p$ | 127 | 124(8) | $p$ |
| $E_g$ | $E$ | $E'$ | $\nu$ (Si–Cl) | $d$ | $d$ | $d$ | 625 | 622(6) | |
| | | | $\delta$ (Si–Cl) | $d$ | $d$ | $d$ | 145 | 132(8) | |
| | | | $\delta$ (Si–Cl) | $d$ | $d$ | $d$ | 202 | 211(8) | $d$ |
| $A_{2u}$ | $A_2$ | $A_2''$ | $\nu$ (Si–Cl) | $i$ | $i$ | $i$ | 478 | | |
| | | | $\delta$ (Si–Cl) | $i$ | $i$ | $i$ | 249 | | |
| $E_u$ | $E$ | $E''$ | $\nu$ (Si–Cl) | $i$ | $d$ | $d$ | 588 | 588(6b) | $d$ |
| | | | $\delta$ (Si–Cl) | $i$ | $d$ | $d$ | 80 | — | |
| | | | $\delta$ (Si–Cl) | $i$ | $d$ | $d$ | 188 | 179(5) | |

$\nu$: stretching vibration

$\delta$: bending vibration

$p$: Raman active and polarized

$d$: Raman active and depolarized

$i$: Raman inactive

It was once suggested that the stable configuration of this molecule is the staggered form. Actually there is no difference in the intensity curve between the staggered and freely rotating model, as long as we confine ourselves to the region of the scattering angle corresponding to $s = 2 \sim 8$. However, in the next outer region at $s = 11$, there has been observed a halo which can be explained only on the basis of the freely rotating model, but not by the staggered form. This is conclusive evidence in favor of the freely rotating model from the electron diffraction experiment. We have already stated that the Raman investigation also rules out the $D_{3d}$ configuration.

We are now interested in the explanation for the difference in the internal rotation between $Cl_3C-CCl_3$ and $Cl_3Si-SiCl_3$. As in the case of the C–C single bond, let us consider that the Si–Si bond itself, i. e. the pair of electrons and the orbital describing their motion, does not resist internal rotation appreciably. Further, let us calculate the interaction between the chlorine atoms not bonded directly by the use of the potential function:

$$V = -\frac{1.515 \times 10^{-10}}{r^6} + \frac{1.620 \times 10^{-7}}{r^{12}} \text{ erg/molecule} \quad (r \text{ in Å})$$

which was obtained by Morino and Miyagawa from the second virial coefficient of chlorine gas.[71] Then we can show at once that the hindering potential is negligible for hexachlorodisilane, while the barrier height amounts to 12 kcal./mole for hexachloroethane in conformity with the experimental value referred to in Section 10. Thus we can explain quite reasonably the difference in the internal rotation between these two molecules. Moreover, through this simple calculation we can see that the steric force plays the most important part in determining the potential restricting internal rotation, as stated in Section 12.

One might consider that the Raman lines of hexachlorodisilane would become broad owing to the free rotation about the Si–Si axis. If, however, there is no appreciable interaction between chlorine atoms bonded to different Si-atoms, the normal frequencies can be shown to be independent of the azimuthal angle of internal rotation.[69] Accordingly the Raman lines would remain fairly sharp even in the case of free rotation.

## Summary of Chapter IV

The structure of other simple molecules with C–C axes of internal rotation studied by similar methods are discussed. Besides these the measurement of optical activity is introduced as another experimental method and is applied to the study of the rotational isomerism of D-secondary butyl alcohol.

In some coordination complexes the ligands of the type, $XH_2C–CH_2X$, have been shown to be in the *gauche* configuration with a mirror image non-superposable. For such complexes more optical iromers exist than would be expected from the classical theory.

The carbon skeleton of cyclopentane is usually considered to have a planar configuration, but this is not compatible with the entropy value of this substance. On the basis of the tetrahedral valence of the carbon atom two forms of the cyclohexane molecule would be possible. However, the thermal, spectroscopic and electron diffraction investigations show that this molecule is predominantly in one form (the chair form) at room temperature. From this form two different configurations of monosubstitution products (polar and equatorial) can be derived. It has been shown that monochlorocyclohexane and methylcyclohexane exist only in the equatorial form in

---

71. Y. Morino and I. Miyagawa, J. Chem. Soc. Japan, **68**, 62 (1947).

which the chlorine atom and the methyl group lie in the general plane of the carbon ring. Similarly in polysubstitution products some configurations are found to be much more stable than the other.

Benzene hexachloride may exist in thirteen different configurations, even if the carbon ring takes only the chair form. Of these thirteen configurations, five are ruled out because of the close approach of chlorine atoms. Of the remaining eight configurations, five are assigned to $\alpha$, $\beta$, $\gamma$, $\delta$ and $\varepsilon$ isomers by the optical and dielectric measurements.

The potential barrier to internal rotation of OH group about the CO axis in methyl and ethyl alcohols, has been found to be about 1000 cal./mole by microwave and thermal measurements. In phenol derivatives, carboxylic acids and esters, the hindrance of rotation of the OH group is largely caused by the partial double bond character of the CO bond. Due to this electronic structure, together with the steric and electrostatic interactions, these molecules exist predominantly in the *cis* form.

In the case of O–O and S–S bonds, the charge density on these atoms is unsymmetrical about the axes and the interaction of these two electronic clouds becomes the dominant factor in determining the azimuth of one rotating group relative to the other, which is approximately 90°.

The Raman and electron diffraction investigations show that the molecule of hexachlorodisilane exerts almost free internal rotation. This is explained as due to the fact that the two $Cl_3Si$ groups are separated at a distance large enough to make the interaction (steric repulsion) between the rotating groups very small.

# CHAPTER V

# Paraffinic Hydrocarbons

## 17. Internal Rotation in Butane, Pentane and Hexane

The simplest hydrocarbon molecule containing a carbon skeleton which can exert internal rotation is n-butane. The spectroscopic behavior of this molecule is expected to be similar to that of the 1,2-dihalogenoethanes in so far as the skeletal vibrations are concerned.

Actually Mizushima, Morino and Nakamura[1] were able to prove that the Raman lines observed in the liquid state can be assigned to the *trans* $(C_{2h})$ and *gauche* forms $(C_2)$ and that only the *trans* lines persist in the solid state, just as in the case of the 1,2-dihalogenoethanes. (See Table 5.2.) Later on these authors, together with Watanabe, Ichishima and Noda, determined the energy difference between these two rotational isomers to be 0.8 kcal./mole from the temperature dependence of the intensity ratio of two Raman lines, one assigned to the *trans* form and the other assigned to the *gauche* form. This agrees with the value adopted by Pitzer with which he was able to calculate a value for the molecular entropy of n-butane in good agreement with the third law value.[2] More exact measurements of energy differences have recently been made by Szasz, Sheppard and Rank who found a value of $\Delta E = 770 \pm 90$ cal./mole.[3] This is in excellent agreement with the value stated above to within the limit of experimental error. The corresponding infrared measurements in the liquid and solid state have been made by Axford and Rank with the same conclusion as was reached from the observation of the Raman effect.[4]

Before starting a discussion of the molecular configurations of more complex hydrocarbons, it is appropriate to introduce notations by which we can represent the various configurations quite simply. Let us propose

---

1. S. Mizushima, Y. Morino and S. Nakamura, Sci. Pap. Inst. Phys. Chem. Res. Tokyo, **37**, 205 (1940).
2. K. S. Pitzer, J. Chem. Phys., **8**, 711 (1940).
3. G. J. Szasz, N. Sheppard and D. H. Rank, J. Chem. Phys., **16**, 704 (1948).
4. D. W. E. Axford and D. H. Rank, J. Chem. Phys. **18**, 51 (1950).

two of them (Shimanouchi and Mizushima).[5] In the first one the directions of C–C bonds are shown in order. If we fix one carbon atom in space and designate the directions of its four bonds by $a$, $b$, $c$, and $d$, then the direction of any other C–C bond contained in the same molecule must be parallel to one of these directions, so long as we confine ourselves to the discussion of saturated hydrocarbon molecules. A unique representation of the carbon skeleton can, therefore, be obtained by denoting the directions of the C–C bonds in order, beginning from one end of the molecule. For example the *trans* form of *n*-butane is represented as $a\ b\ a$, one of the *gauche* forms as $a\ b\ c$,

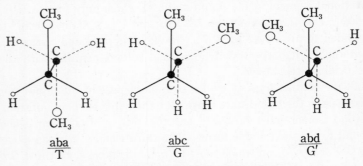

Fig. 5.1. Stable configurations of *n*-butane.

and the other *gauche* form as $a\ b\ d$. If there is a side chain attached to the normal chain, it is denoted by a parenthesis: for example, isobutane $CH_3$–$CH(CH_3)$–$CH_3$ is represented as $a\ (c)\ b$. In the second notation we denote the carbon skeleton by the permutation of the *trans* ($T$), one *gauche* ($G$) and the other *gauche* ($G'$) forms of *n*-butane shown in Fig. 5.1. For example, the five conceivable configurations of *n*-pentane can be represented by this notation as:

$$T\,T \qquad T\,G \qquad T\,G' \qquad G\,G \qquad G'\,G'$$

which in turn are represented by the first notation as:

$$a\,b\,a\,b \qquad a\,b\,a\,c \qquad a\,b\,a\,d \qquad a\,b\,c\,d \qquad a\,b\,d\,c.$$

The rotational isomers which contain $G\,G'$ and $G'\,G$ configurations are not included in this series, since they must be very unstable because of the close approach of non-bonded atoms. By the second notation we can easily determine the symmetry properties (e. g. the configurations represented by $T\,\dot{T}, G\,G$, and $G'\,G'$ have twofold axes of symmetry) and mirror images (e. g.

---

5. T. Shimanouchi and S. Mizushima, Sci. Pap. Inst. Phys. Chem. Res. Tokyo, **40,** 467 (1943).

$T\,G'$ and $G'\,G'$ are the mirror images of $T\,G$ and $G\,G$, respectively), or we can at once enumerate all the conceivable isomers. This is much better than the first notation in so far as the normal paraffins are concerned. For *iso*-paraffins the situation is somewhat different, but we can still use the second notation by adding suitable subsymbols (see Table 5.1).

Mizushima, Morino and Takeda measured the Raman spectra of *n*-pentane in the liquid and solid states and found a marked difference between the two spectra.[6] (See Table 5.2.) From the number of lines observed in the liquid and solid states it can be concluded that there must be at least two rotational isomers in the liquid state, of which only one persists in the solid state. This has been shown by Mizushima and Shimanouchi[7] to have the extended form ($a\,b\,a\,b$ or $T\,T$). Rank, Sheppard and Szasz have also obtained the Raman spectra of liquid and solid *n*-pentane with the same conclusion as stated above.[8]

As all the Raman lines of solid *n*-pentane persist in the liquid state, one of the isomers in the liquid state must have the extended configuration ($T\,T$). The second isomer most probably has the $T\,G$ or $T\,G'$ form, since these correspond to the next stable forms. The third isomer, if it can exist at all, has the $G\,G$ or $G'\,G'$ form which must be more unstable than the $T\,G$ or $T\,G'$ isomer, since the third isomer contains more *gauche* forms. From the number of lines observed in the liquid state it seems that, at least at ordinary temperatures, we can treat liquid *n*-pentane as the mixture of two isomeric forms.

Mizushima and Okazaki[9] observed the temperature-dependence of the intensity of the two Raman lines at 864 and 838 cm.$^{-1}$ which are assigned, respectively, to the first and the second configurations, and obtained a value of the energy difference of 0.53 kcal./mole in good agreement with the result of Sheppard and Szasz.[10]

*n*-Hexane contains three axes of internal rotation so that the conceivable number of rotational isomers becomes much larger than for *n*-pentane. In Table 5.1 are shown the configurations of the rotational isomers represented by the two notations referred to above together with those of 2-methylpentane, 3-methylpentane, 2,3-dimethylbutane and 2,2-dimethylbutane, all corre-

---

6.  S. Mizushima, Y. Morino and M. Takeda, Sci. Pap. Inst. Phys. Chem. Res. Tokyo, **38**, 437 (1941).

7.  S. Mizushima and T. Shimanouchi, Proc. Imp. Acad. Tokyo, **20**, 86 (1944).

8.  D. N. Rank, N. Sheppard and G. J. Szasz, J. Chem. Phys., **17**, 83 (1950).

9.  S. Mizushima and H. Okazaki, J. Am. Chem. Soc., **71**, 3411 (1949).

10.  N. Sheppard and G. J. Szasz, J. Chem. Phys., **17**, 86 (1949).

### TABLE 5.1
Rotational isomers of hexane

| Skeleton | Notation I | Notation II | Weight | Symmetry number |
|---|---|---|---|---|
| C–C–C–C–C–C | $a\ b\ a\ b\ a$ | $T\ T\ T$ | 1 | 2 |
| | $a\ b\ a\ b\ c$ | $T\ T\ G$ | 2 | 1 |
| | $a\ b\ a\ b\ d$ | $T\ T\ G'$ | 2 | 1 |
| | $a\ b\ a\ c\ a$ | $T\ G\ T$ | 1 | 2 |
| | $a\ b\ a\ d\ a$ | $T\ G'\ T$ | 1 | 2 |
| | $a\ b\ a\ c\ d$ | $T\ G\ G$ | 2 | 1 |
| | $a\ b\ a\ d\ c$ | $T\ G'\ G'$ | 2 | 1 |
| | $a\ b\ c\ b\ a$ | $G\ T\ G'$ | 2 | 1 |
| | $a\ b\ c\ b\ d$ | $G\ T\ G$ | 1 | 2 |
| | $a\ b\ d\ b\ c$ | $G'\ T\ G'$ | 1 | 2 |
| | $a\ b\ c\ d\ a$ | $G\ G\ G$ | 1 | 2 |
| | $a\ b\ d\ c\ a$ | $G'\ G'\ G'$ | 1 | 2 |
| C–C–C–C–C<br>\|<br>C | $a\ (c)\ b\ a\ b$ | $(T + G')\ T$ | 1 | 1 |
| | $a\ (d)\ b\ a\ b$ | $(T + G)\ T$ | 1 | 1 |
| | $a\ (c)\ b\ a\ d$ | $(T + G')\ G'$ | 1 | 1 |
| | $a\ (d)\ b\ a\ c$ | $(T + G)\ G$ | 1 | 1 |
| | $a\ (c)\ b\ d\ b$ | $(G + G')\ T$ | 1 | 1 |
| C–C–C–C–C<br>\|<br>C | $a\ b\ (c)\ a\ b$ | $(T + G)\ (G' + T)^*$ $\overleftarrow{\phantom{x}}\ \overrightarrow{\phantom{x}}$ | 1 | 1 |
| | $a\ b\ (c)\ a\ c$ | $(T + G)\ (T + G)$ $\overleftarrow{\phantom{x}}\ \overrightarrow{\phantom{x}}$ | 1 | 1 |
| | $a\ b\ (d)\ a\ d$ | $(T + G')\ (T + G')$ $\overleftarrow{\phantom{x}}\ \overrightarrow{\phantom{x}}$ | 1 | 1 |
| | $a\ b\ (c)\ a\ d$ | $(T + G)\ (G + G')$ $\overleftarrow{\phantom{x}}\ \overrightarrow{\phantom{x}}$ | 1 | 1 |
| | $a\ b\ (d)\ a\ c$ | $(T + G')\ (G' + G)$ $\overleftarrow{\phantom{x}}\ \overrightarrow{\phantom{x}}$ | 1 | 1 |
| | $a\ b\ (d)\ c\ d$ | $(G + G')\ (T + G)$ $\overleftarrow{\phantom{x}}\ \overrightarrow{\phantom{x}}$ | 1 | 1 |
| | $a\ b\ (c)\ d\ c$ | $(G' + G)\ (T + G')$ $\overleftarrow{\phantom{x}}\ \overrightarrow{\phantom{x}}$ | 1 | 1 |
| C–C–C–C<br>\|  \|<br>C  C | $a\ (c)\ b\ (a)\ c$ | $T + G + T + G'$ | 1 | 2 |
| | $a\ (c)\ b\ (d)\ a$ | $G + G' + T + G'$ | 1 | 2 |
| | $a\ (d)\ b\ (c)\ a$ | $G' + G + T + G$ | 1 | 2 |
| C<br>\|<br>C–C–C–C<br>\|<br>C | $a\ (c)\ (d)\ b\ a$ | $T + G + G'\cdot$ | 3 | 1 |

* The arrow indicates that the operation should be made only in that direction.

sponding to the molecular formula of $C_6H_{14}$. Subsymbols have been used to represent configurations of branched hydrocarbons by the second notation. (Shimanouchi and Mizushima.)[5]

Mizushima, Morino and Takeda[6] observed the Raman spectra of $n$-hexane at room temperature and at very low temperatures and found that the spectrum of liquid $n$-hexane undergoes a marked simplification upon freezing. (See Table 5.2.) All of the lines from the less stable isomers disappear, leaving only the lines from the most stable isomer. This has been shown to have the extended form $(TTT)$ by Mizushima and Shimanouchi. This experimental result is in good agreement with that of Sheppard and Szasz[10] recently published and is also supported by the low temperature infrared measurements of Axford and Rank.[4]

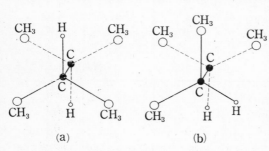

(a)                                (b)

Fig. 5.2. Stable configurations of
2,3-dimethylbutane in the liquid state.

It has been reported by Szasz and Sheppard[11] and Axford and Rank[4] that the Raman and infrared spectra of 2,3-dimethylbutane did not change significantly when the temperature was lowered. Also all of the Raman and infrared lines were found to persist in the solid state. Therefore, this seemed to form a contrast to the straight hydrocarbons for which the crystallization of the liquid produce a very marked simplification of their Raman and infrared spectra. The behavior of 2,3-dimethylbutane may be acounted for either by (a) the rotational isomers having nearly equal energies, or (b) one rotational isomer having a much higher energy than the other so that virtually only one form exists in the liquid and solid states.

Brown and Sheppard[12] have made more careful measurements on this substance and have shown that the rapid freezing of the liquid often leads to a glassy solid having a spectrum virtually identical with that of the liquid. However, in the really crystalline state the infrared spectrum of 2,3-dimethyl-butane was greatly simplified, a number of strong absorptions disappearing, and the spectrum indicated that the configuration stable in the lattice has a center of symmetry. This is, therefore, the *trans* form (form *a* of Fig. 5.2).

11.　G. J. Szasz and N. Sheppard, J. Chem. Phys., 17, 93 (1949).
12.　J. K. Brown and N. Sheppard, J. Chem. Phys., 19, 976 (1951).

The other form (form $b$) is the *gauche* form which can be obtained from the *trans* form by an internal rotation through an angle of about 120°. Evidently the *gauche* form has a mirror image and its weight factor is two. There is thus little doubt that the interpretation of almost zero energy difference between the two isomers (*trans* and *gauche*) in the liquid state is a correct one.

## 18. Internal Rotation in Long Chain $n$-Paraffins

Mizushima, Morino, Okazaki and others have measured the Raman spectra of the $n$-paraffins from $C_4$ to $C_{12}$ and $C_{16}$ in the liquid and the solid states and have found in all of them a pronounced simplification of the spectra upon freezing as shown in Table 5.2.[6, 9, 13, 14] From the analogy to the molecular configuration of the more simple $n$-paraffins, it may be concluded that all these $n$-paraffins have extended molecular configurations $(TTT..T)$ in the solid state. This conclusion is in agreement with the result of normal vibration calculation made by Mizushima and Shimanouchi (see next section) and is compatible with the experimental results of X-ray diffraction by Müller on long chain $n$-paraffins.[15]

Before entering into the mathematical discussion of the normal vibrations of long chain molecules, let us treat the chain vibrations on the basis of the rod model, so that we may picture at least a part of the problem (Shimanouchi and Mizushima).[5]

If we approximate the extended carbon chain by a continuous rod, then the frequency $\nu$ of its longitudinal motion will be given by

$$\nu = \frac{1}{2 l} \sqrt{\frac{E}{d}} \tag{5.1}$$

where $E$ is Young's modulus, $d$ the density and $l$ the length of the rod. This frequency is inversely proportional to the length of the rod and accordingly, we may expect the appearance of a Raman frequency which is inversely proportional to the number of carbon atoms.

Actually we observe such a Raman line for each of the $n$-paraffins in the lower frequency region as shown in Table 5.2. This is the only line observed in the solid state below 500 cm.$^{-1}$ and its frequency is inversely proportional to the number of carbon atoms, if the molecule contains more than five carbon atoms.

13. S. Mizushima, Y. Morino and S. Nakamura, Sci. Pap. Inst. Phys. Chem. Res. Tokyo, **37**, 205 (1940).

14. S. Mizushima and T. Shimanouchi, J. Am. Chem. Soc., **71**, 1320 (1949).

15. A. Müller, Proc. Roy. Soc., **A 124**, 317 (1927), **125**, 437 (1928), **127**, 417 (1930).

We may obtain the value of Young's modulus from Equation (5.1) by putting the observed value of the Raman frequency into $\nu$ on the left-hand side. This has been found to be $34 \times 10^{11}$ dynes/cm.[2] and accordingly, the carbon chain has the modulus which is of the same order of magnitude as diamond. This is quite reasonable, because we are not concerned with the bulk Young's modulus of ordinary solid paraffin, but we deal here with the Young's modulus of the paraffin molecule itself which is connected by covalent bonds and which, therefore, should not be much different from that of diamond. From this result we see that the vibrational model referred to above is quite appropriate.

TABLE 5.2

Raman spectra of $n$-paraffins in the liquid and solid states

| $n$-Butane | | $n$-Pentane | | $n$-Hexane | |
|---|---|---|---|---|---|
| Liquid | Solid | Liquid | Solid | Liquid | Solid |
| 207(00) | | | | | |
| 223(0) | | | | | |
| 259(0) | | | | | |
| 287(0) | | | | | |
| | | | | 315(1 b) | |
| 320(0) | | 333(1) | | 335(1) | |
| | | | | 370(4) | 373(3) |
| | | 400(7) | 406(3) | 401(2) | |
| 429(5) | 425(4) | | | | |
| | | 467(2) | | 450(1) | |
| | | | | 746(0) | |
| | | 764(3) | | 760(0) | |
| 789(2) | | | | 794(0) | |
| 809(0) | | | | 810(1) | |
| 827(6) | | | | 826(3) | |
| 837(7) | 837(6) | 838(5) | | | |
| | | 864(4) | 869(3) | 868(3) | |
| | | | | 890(4) | |
| | | 904(1 b) | | 900(2) | 898(3) |
| 955(1 b) | | 953(0) | | 952(0) | |
| | | | | 976(00) | |
| 980(2) | | 990(1) | | | |
| | | | | 1005(1) | 1005(0) |
| | | 1025(3) | | | |
| | | 1035(3) | 1031(0) | 1040(3) | |
| 1057(4) | 1059(5) | | | 1065(1) | 1064(3) |
| 1077(1) | | 1072(3 b) | 1069(3) | | |
| | | | | 1080(4) | |

| n-Butane | | n-Pentane | | n-Hexane | |
|---|---|---|---|---|---|
| Liquid | Solid | Liquid | Solid | Liquid | Solid |
| 1150(2) | 1151(4) | 1142(3) | 1145(3) | 1140(3) | 1143(3) |
| 1168(0) | | 1165(0) | | 1165(0) | |
| | | | | 1220(00) | |
| | | 1264(1) | | 1250(0) | |
| 1281(0) | | | | | |
| 1301(1b) | 1300(4) | 1302(4b) | 1303(3) | 1302(5b) | 1300(4) |
| | | | | 1343(00) | |
| | | | | 1366(0) | |
| 1444(5b) | 1442(3) | 1437(5b) | 1445(4) | 1436(5b) | 1450(5) |
| | | 1458(6b) | 1466(5) | 1459(7b) | 1462(4) |
| 2666(1) | | 2666(1) | 2666(1) | 2672(1) | |
| 2702(1) | 2703(1) | 2713(1) | | 2703(0) | 2696(1) |
| 2733(3) | 2725(1) | 2733(2) | 2720(2) | 2735(2) | 2723(2) |
| | 2853(8) | 2847(5) | 2848(2) | 2849(8) | 2851(5) |
| 2860(8) | | 2861(5) | 2859(2) | 2862(5) | |
| 2877(10) | 2872(8) | 2875(10) | | 2875(10) | 2871(5) |
| | | | 2885(8) | | 2885(8) |
| 2901(2) | 2896(10) | | | | |
| 2914(5) | 2912(4) | 2908(5b) | 2909(4) | 2916(5) | 2907(5) |
| 2938(8b) | 2931(1) | 2936(8b) | 2934(5) | 2939(10) | 2934(5) |
| | 2950(4) | | 2956(2) | | 2953(3) |
| 2962(6b) | 2965(9) | 2964(8b) | 2966(6) | 2964(10b) | 2965(6) |

| n-Heptane | | n-Octane | | n-Nonane | |
|---|---|---|---|---|---|
| Liquid | Solid | Liquid | Solid | Liquid | Sclid |
| 198(0) | | 196(0) | | 196(0) | |
| 222(0) | | 218(00) | | 221(0) | |
| | | 241(00) | | 248(3) | 249(2) |
| 285(1) | | 279(5) | 283(3) | 264(4) | |
| 310(6) | 311(5) | 295(1) | | 283(1) | |
| | | 347(0) | | 340(1) | |
| 358(2) | | 374(1/2) | | 375(1/2) | |
| | | 399(0) | | 404(1) | |
| 394(3) | | 427(1) | | 417(1/2) | |
| 404(0) | | 454(0) | | 453(1/2b) | |
| | | | | 489(0) | |
| 455(1) | | 505(0b) | | 510(0) | |
| 506(2) | | | | 523(0) | |
| 695(0) | | | | 548(0) | |
| | | 696(00) | | 697(0) | |
| 721(1) | | 723(1/2) | | 721(1/2) | |
| 742(1) | | 735(1/2) | | 751(1/2) | |
| 775(2b) | | 765(1) | | | |

| n-Heptane | | n-Octane | | n-Nonane | |
|---|---|---|---|---|---|
| Liquid | Solid | Liquid | Solid | Liquid | Solid |
| 807(0) | | | | 781(1) | |
| 829(2) | | | | | |
| 838(6) | | 815(2b) | | 820(3) | |
| 851(3) | | 843(2) | | 843(3) | |
| 866(0) | | 861(3) | | 871(3) | |
| 888(2) | | 878(3) | | | |
| 901(5) | | 896(4) | 899(2) | 892(4) | 888(2) |
| 909(3) | 905(5) | | | | |
| 930(1/2) | | | | 926(0) | |
| 950(0) | | 953(1) | | | |
| 960(1) | | 970(2) | | 970(1 b) | |
| 988(0) | | 999(00) | | | |
| | | | | | |
| 1023(1 b) | 1028(1) | 1026(2) | | 1019(1) | |
| 1044(3) | | 1045(2) | 1045(1) | 1046(0) | |
| 1057(2) | 1056(3) | 1060(4) | 1062(2) | 1062(3) | 1060(1) |
| 1071(2) | 1072(2) | 1083(4 b) | | 1082(3 b) | |
| 1086(4 b) | | | | | |
| 1139(4) | 1139(4) | 1137(4) | 1138(2) | 1134(3) | 1136(2) |
| 1163(2) | | 1162(1) | 1176(0) | 1159(1) | |
| 1206(0) | | 1199(0) | | 1192(0) | |
| 1237(1) | | 1226(0) | | 1217(0) | |
| 1264(0) | | | | | |
| 1281(0) | | | | | |
| 1298(6 b) | 1296(4) | 1299(5 b) | 1297(3) | 1299(4 b) | 1297(3) |
| 1312(2) | | | | | |
| 1342(1) | | 1342(0) | | 1340(0) | |
| 1364(0) | | 1366(0 b) | 1384(1) | 1365(0) | |
| 1432(7 b) | 1446(4 b) | 1434(7 b) | 1449(3 b) | 1433(8) | 1444(2) |
| 1460(7 b) | 1473(3) | 1459(5 b) | 1469(3 b) | 1460(8) | 1465(3) |
| | | | | | 1486(0) |
| | | | | | |
| 2670(1) | | 2668(1) | | 2666(1) | 2660(1) |
| 2708(1) | | 2714(0) | 2699(0) | 2706(1) | |
| 2730(3) | 2725(1) | 2732(3) | 2727(0) | 2731(2) | 2720(0) |
| 2848(10 b) | 2852(3) | | 2765(0) | | |
| | 2866(2 b) | 2849(10) | 2853(3 b) | 2846(10) | 2847(2) |
| 2873(8) | 2878(7) | 2873(6) | 2882(10) | 2873(2) | 2877(5) |
| | | 2904(6) | | 2896(3) | |
| 2901(5 b) | 2904(2) | 2937(8) | 2937(2) | 2933(7) | 2934(2) |
| | | 2965(8 b) | 2965(5) | 2964(8) | 2964(4) |
| 2937(8) | 2930(3) | | | | |
| | 2950(2) | | | | |
| 2963(10 b) | 2964(7) | | | | |

| n-Decane | | n-Dodecane | | n-Cetane | |
|---|---|---|---|---|---|
| Liquid | Solid | Liquid | Solid | Liquid | Solid |
| | | | | | 150(3) |
| | | 195(1) | 194(2) | 199(1) | |
| 200(0) | | 218(1) | | 215(2) | |
| 230(3) | 231(3) | 239(2b) | | 231(1) | |
| 250(4) | | | | 278(0b) | |
| | | 308(0) | | 330(0) | |
| 339(0) | | 349(0) | | 356(0) | |
| 359(1/2) | | | | | |
| 404(2b) | | 397(1) | | 404(1) | |
| | | 421(1/2) | | | |
| 441(1/2) | | 453(00) | | 450(00) | |
| 465(1/2) | | 487(00) | | | |
| 521(1) | | | | | |
| 661(0b) | | | | | |
| 697(1/2) | | | | | |
| 723(1) | | 724(1) | | 724(1) | |
| 744(1) | | 749(0) | | 742(00) | |
| 772(2b) | | 772(1b) | | 762(00) | |
| | | | | 784(00) | |
| 810(2b) | | 805(1) | | 808(0) | |
| | | 817(1) | | | |
| 844(3b) | | 845(2b) | | 839(0) | |
| 870(1) | | 872(1) | | 871(2) | |
| 886(3) | 886(2) | | | | |
| 898(3) | | 893(3) | 892(1) | 894(3) | 888(1) |
| 921(1) | | 917(00) | | | |
| | | 931(00) | | | |
| 952(1) | | 955(1) | | | |
| 971(1) | | 964(1) | | 962(1) | |
| 991(1) | | | | 997(00) | |
| 1008(1) | | 1002(00) | | | |
| 1023(1) | | 1032(1b) | | 1015(0) | |
| 1047(2) | | | | | |
| 1062(4) | 1060(3) | 1061(3) | 1061(2) | 1063(4) | 1058(2) |
| 1080(4) | | 1078(3) | | 1082(5b) | |
| 1092(3) | | 1100(0) | | 1108(0) | |
| 1133(4) | 1136(3) | 1129(3) | 1136(3) | 1132(4) | 1135(2) |
| 1161(2) | | 1159(1) | | 1165(0) | |
| 1189(0) | | | | | |
| 1211(0) | | | | | |
| 1250(0) | | | | | |
| 1271(0) | | | | | |
| 1301(6b) | 1295(3) | 1300(6b) | 1297(3) | 1304(7b) | 1295(2) |

| n-Decane | | n-Dodecane | | n-Cetane | |
|---|---|---|---|---|---|
| Liquid | Solid | Liquid | Solid | Liquid | Solid |
| | | | | 1314(2) | |
| 1340(1/2) | | 1341(1/2) | | 1344(0) | |
| 1366(1) | | 1369(1) | | 1368(1) | |
| 1433(8 b) | 1447(2 b) | 1433(7 b) | 1441(3 b) | 1435(8 b) | 1442(2 b) |
| 1460(6 b) | 1475(1) | 1461(5 b) | 1462(2 b) | 1460(6 b) | 1471(2 b) |
| | | | | | |
| 2673(1) | 2660(1) | 2670(0 b) | | 2673(0 b) | |
| | | 2718(0) | | 2713(2) | |
| 2732(3) | 2724(1) | 2730(3 b) | 2723(0 b) | 2730(3) | |
| 2850(10) | 2843(2) | 2849(10) | 2845(4 b) | 2847(10 b) | 2846(3) |
| | 2877(6) | | 2879(10) | 2885(8 b) | 2878(5) |
| 2897(6 b) | | 2896(8 b) | | | |
| 2936(7) | 2935(1) | 2936(8) | 2933(1) | 2937(8 b) | 2934(1) |
| 2966(8) | 2964(3) | 2964(9 b) | 2964(3) | 2965(6 b) | 2963(2) |

We can show by the normal vibration calculation to be described below that this frequency corresponds to a deformation vibration of the extended chain, in which the change of the polarizability is marked, and accordingly it can appear as a strong Raman line. Now since this frequency changes considerably with the number of carbon atoms, we can use this for the determination of the number of carbon atoms in the chain, in so far as the spectrum in the solid state is concerned.

In the liquid state, however, there appear many other Raman lines in the lower frequency region as shown in Table 5.2. This means that there are several rotational isomers in this state. In the homologous series from butane $(C_4)$ to dodecane $(C_{12})$ one of these rotational isomers should be of the extended configuration $(T T \ldots T)$, because the line observed in the solid state persists in the liquid state. However, in cetane the characteristic line of the solid at 150 cm.$^{-1}$ has escaped detection in the spectrum of the liquid and accordingly we have to conclude that the extended configuration of cetane molecule does not exist in an appreciable amount in the liquid state. This might have to do with the idea proposed by Eyring et al. who have shown that as hydrocarbon chains increase in length, they do not flow as a single unit, but tend to flow as segments.[16]

In the case of cetane in the liquid state the Raman lines observed in the lower frequency region have the frequencies as shown in Table 5.3. According

---

16. W. Kauzmann and H. Eyring, J. Am. Chem. Soc., **62**, 3113 (1940).

TABLE 5.3

The Raman lines of liquid $n$-cetane in the lower frequency region

| Raman lines | Number of carbon atoms |
|---|---|
| 199(1) | 12 |
| 215(2) | 11 |
| 231(1) | 10 |
| 278(0) | 9 |
| 330(0) | 7 |
| 356(0) | 6 |
| 404(1) | 5 |
| 450(0) | 4 |

to what we have just stated, these Raman lines should be considered to arise from the normal vibrations of folded forms. Let us tentatively assume that each of these frequencies corresponds to the deformation vibration of a segment of the chain molecule which can exert this vibration irrespective of the motion of other part of the molecule. We can then assign to each frequency the number of carbon atoms of the segment as shown in Table 5.3 (Shimanouchi and Mizushima).[5] It is very interesting that these numbers can be grouped in such a way that the sum of two numbers is always sixteen:

$$12 + 4 = 16$$
$$11 + 5 = 16$$
$$10 + 6 = 16$$
$$9 + 7 = 16$$

Here the two lines in each pair have almost the same intensity. Therefore, we may consider that the rotational isomers of liquid $n$-cetane have the following configurations. (See Fig. 5.3.)

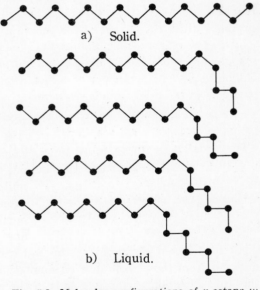

a)   Solid.

b)   Liquid.

Fig. 5.3. Molecular configurations of $n$-cetane in the liquid and solid states.

$$T\,T\,T\,T\,T\,T\,T\,T\,T\,G\,T\,T$$
$$T\,T\,T\,T\,T\,T\,T\,T\,G\,T\,T\,T$$
$$T\,T\,T\,T\,T\,T\,T\,G\,T\,T\,T\,T$$
$$T\,T\,T\,T\,T\,T\,G\,T\,T\,T\,T\,T$$

We do not know whether such a simple assumption really conforms to the facts, but it may be useful for the further development of the study of configurations of long chain molecules in the liquid state. At any rate we are sure that there are different folded forms for long chain paraffins in the liquid state and that the transition from a single extended form of the solid into various configurations of the liquid must give rise to a part of the entropy increase on fusion.

We know that there is an alternation in melting point as we go up the homologous series of $n$-paraffins (Fig. 5.4). The melting point $T$ is related to the heat of fusion $\Delta H$ and the entropy change $\Delta S$ by

$$T = \frac{\Delta H}{\Delta S} \tag{5.2}$$

Therefore, the alternation of melting point is due to the alternation in $\Delta H$ or in $\Delta S$. The result of calculation shows that the former one is more important. This is quite understandable, because the extended forms with an odd number of carbons in which the lines joining the two end carbons point in different directions at the two ends of the chain, are definitely different from those with an even number in which the end lines point in the same direction at the two ends of the chain. (See Fig. 5.5.)

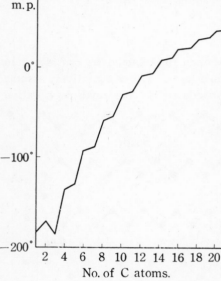

Fig. 5.4. Melting points of $n$-paraffins as a function of number of carbon atoms in the chain.

We know further that the melting point converges to a limiting temperature as the number of carbon atoms increases. On the other hand the heat of fusion can be expressed by a linear equation of the type:

$$\Delta H = q_0 + q_1 N \tag{5.3}$$

where $N$ is the number of carbon atoms (Huggins).[17] If we assume that $\Delta S$ does not change much with the number of carbon atoms, the melting point must be raised linearly with the number of carbon atoms. Therefore, in order that the melting point may converge to a limiting temperature, $\Delta S$ should contain a term which is proportional to the number of carbon atoms. This term must evidently arise from the increase of the number of rotational isomers on fusion. If we assume approximately that the three potential minima in the internal rotation about a carbon single bond have almost the same energy, the entropy increase on fusion can be expressed as

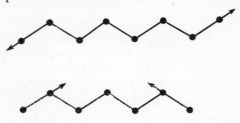

Fig. 5.5. An odd and an even numbered chain molecule. $CH_2$ groups are marked by black circles.

$$\Delta S = S_0 + k N \ln 3 \quad (5.4)$$

where $k$ is the Boltzmann constant and $S_0$ the entropy increase due to the mobility of the molecules against each other which is independent of the number of carbon atoms. Therefore, the convergent point can be calculated as

$$T = \frac{\Delta H}{\Delta S} = \frac{q_1}{k \ln 3} = 330° \text{ K} \quad (5.5)$$

This is not much different from the observed value of 370° K. The small difference between the theoretical and experimental values may be due to the fact that the three potential minima have not the same energy (Mizushima and Shimanouchi).[18]

## 19. Skeletal Vibrations of $n$-Paraffins

In the preceding section the vibrational problem of paraffin molecules has been treated by a rough approximation and it is desirable at this stage to make a more exact explanation of this problem. For this purpose let us calculate the skeletal vibrations of $n$-paraffin chains and assign at least a part of the observed frequencies. The assignment of the other frequencies will be made in Section 7, Part II, taking into account the motions of hydrogen atoms.

17. M. L. Huggins, J. Phys. Chem., **43**, 1083 (1939).
18. S. Mizushima and T. Shimanouchi, Advances in Quantum Physics II, 163 (1947).

As we shall show in Part II, a reasonable form of the potential function which we need in our calculation of normal vibrations is the Urey-Bradley form which can be expressed for $n$-butane and $n$-pentane as follows:

$$V = K_2' r_0 (\Delta r_1 + \Delta r_{N-1}) + \frac{1}{2} K_2 [(\Delta r_1)^2 + (\Delta r_{N-1})^2]$$

$$+ K_1' r_0 \sum_{n=2}^{N-2} (\Delta r_n) + \frac{1}{2} K_1 \sum_{n=2}^{N-2} (\Delta r_n)^2 + H' r_0 \sum_{n=1}^{N-2} (r_0 \Delta \alpha_n) +$$

$$+ \frac{1}{2} H \sum_{n=1}^{N-2} (r_0 \Delta \alpha_n)^2 + F' q_0 \sum_{n=1}^{N-2} (\Delta q_n) + \frac{1}{2} F \sum_{n=1}^{N-2} (\Delta q_n)^2 \qquad (5.6)$$

where $K_1'$ and $K_1$ are the force constants of the C – C bonds except those of the end groups which are designated as $K_2'$ and $K_2$, and $F'$ and $F$ are the force constants between the non-bonded carbon-carbon atoms and $H'$ and $H$ are those for the deformation of valence angles. $r_0$, which is the equilibrium distance of the carbon single bond, is introduced in order to make all the force constants dimensionally similar. The significance of the increments in atomic distances $\Delta r$ and $\Delta q$ and in bond angles $\Delta \alpha$ will be clear from Fig. 5.6.

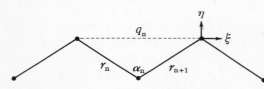

Fig. 5.6.   Skeleton of $n$-pentane.

The calculation of normal vibrations is made according to the process to be described in Chapter II of Part II, putting $N = 4$ for $n$-butane and $N = 5$ for $n$-pentane and using the following values of force constants:

$$\left.\begin{array}{l} K_1 = 4.0 \ \times 10^5 \ \text{dynes/cm.} \\ K_2 = 3.2 \ \times 10^5 \ \text{dynes/cm.} \\ H \ = 0.11 \times 10^5 \ \text{dynes/cm.} \\ F \ = 0.96 \times 10^5 \ \text{dynes/cm.} \\ F' = 0 \\ \alpha_0 \ = \text{tetrahedral angle.} \end{array}\right\} \qquad (5.7)$$

In Table 5.4 are shown the frequencies calculated for the symmetric vibrations for $n$-butane and $n$-pentane which are in good agreement with those observed in the Raman effect. Accordingly these values of force constants can be considered to be of reasonable magnitude.

The calculation of the skeletal vibrations of long chain molecules was made by several authors. In the following the formulation of Kirkwood[19] will be used with, however, a modification made by Shimanouchi and Mizushima[5,7] in the potential function, since the simple valence force model (see Part II) cannot account for quantitative feature of the problem. The potential function to be used in the following calculation is of the same type as equation (5.6) in which, however, we need not distinguish $K_2'$ and $K_2$ from $K_1'$ and $K_1$, since we are dealing with the vibration of a finite segment of an infinite chain. The equations of motion become:

$$\ddot{\xi}_n + \mu \left[ K_1 + 2F'(1 + \cos\alpha_0) \right] \left[ \sin^2\left(\frac{\alpha_0}{2}\right)(2\xi_n - \xi_{n+1} - \xi_{n-1}) \right.$$

$$\left. + (-1)^n \sin\frac{\alpha_0}{2}\cos\frac{\alpha_0}{2}(\eta_{n+1} - \eta_{n-1}) \right] + \mu H \left[ \cos^2\left(\frac{\alpha_0}{2}\right)(2\xi_n - \xi_{n+2} - \xi_{n-2}) \right.$$

$$\left. + (-1)^n \sin\frac{\alpha_0}{2}\cos\frac{\alpha_0}{2}(2\eta_{n-1} - 2\eta_{n+1} + \eta_{n+2} - \eta_{n-2}) \right]$$

$$+ \mu F \left[ 2\xi_n - \xi_{n+2} - \xi_{n-2} \right] = 0. \tag{5.8}$$

$$\ddot{\eta}_n + \mu \left[ K_1 + 2F'(1 + \cos\alpha_0) \right] \left[ \cos^2\left(\frac{\alpha_0}{2}\right)(2\eta_n - \eta_{n+1} - \eta_{n-1}) \right.$$

$$\left. + (-1)^n \sin\frac{\alpha_0}{2}\cos\frac{\alpha_0}{2}(\xi_{n+1} - \xi_{n-1}) \right]$$

$$+ \mu H \left[ \sin^2\left(\frac{\alpha_0}{2}\right)(6\eta_n - 4\eta_{n+1} - 4\eta_{n-1} + \eta_{n+2} - \eta_{n-2}) \right.$$

$$\left. - (-1)^n \sin\frac{\alpha_0}{2}\cos\frac{\alpha_0}{2}(2\xi_{n+1} - 2\xi_{n-1} + \xi_{n+2} - \xi_{n-2}) \right] \tag{5.9}$$

$$- \mu F' \left[ 6\eta_n - 4\eta_{n+1} - 4\eta_{n-1} + \eta_{n+2} + \eta_{n-2} \right] = 0,$$

where $\mu$ is the reciprocal of the mass of the $CH_2$-group, $\xi$ the displacement from equilibrium along the length of the chain and $\eta$ the displacement perpendicular to the chain. (See Fig. 5.1.) For a paraffin molecule with $N$ carbon atoms there are $2N$ such equations of motion which have particular solutions of the form:

$$\xi_n = A e^{i(\omega t + n\lambda)}, \qquad \eta_n = (-1)^n B e^{i(\omega t + n\lambda)}.$$

19. J. G. Kirkwood, J. Chem. Phys., 7, 506 (1939).

TABLE 5.4

The symmetric frequencies of $n$-butane and $n$-pentane molecules in the extended form (Mizushima and Shimanouchi)[5, 7]

| Substance | Frequency in cm.$^{-1}$ | |
| --- | --- | --- |
| | Calculated | Observed |
| $n$-butane | 432 | 425 |
| | 835 | 837 |
| | 1075 | 1059 |
| $n$-pentane | 205 | — |
| | 402 | 406 |
| | 851 | 869 |
| | 1048 | 1069 |

where the amplitudes $A$ and $B$ satisfy the following equations:

$$\left.\begin{array}{c} (\varkappa_{11} - \omega^2)\, A + i\, \varkappa_{12}\, B = 0 \\ -\, i\, \varkappa_{12}\, A + (\varkappa_{22} - \omega^2)\, B = 0 \end{array}\right\} \tag{5.10}$$

$$\varkappa_{11} = \mu\, (1 - \cos \lambda)\, [\{K_1 + 2\, F'\, (1 + \cos \alpha_0)\}\, (1 - \cos \alpha_0) \\ + 2\, H\, (1 + \cos \alpha_0)\, (1 + \cos \lambda) + 4\, F\, (1 + \cos \lambda)]$$

$$\varkappa_{22} = \mu\, (1 + \cos \lambda)\, [\{K_1 + 2\, F'\, (1 + \cos \alpha_0)\}\, (1 + \cos \alpha_0) \\ + 2\, H\, (1 - \cos \alpha_0)\, (1 + \cos \lambda) - 4\, F'\, (1 + \cos \lambda)]$$

$$\varkappa_{12} = \mu \sin \lambda\, [- \{K_1 + 2\, F'\, (1 + \cos \alpha_0)\} \sin \alpha_0 + 2\, H \sin \alpha_0\, (1 + \cos \lambda)]$$

The values of $\omega^2$ corresponding to non-trivial solutions of Equation (5.10) satisfy the secular equation:

$$\left| \begin{array}{cc} \varkappa_{11} - \omega^2 & i\, \varkappa_{12} \\ -\, i\, \varkappa_{12} & \varkappa_{22} - \omega^2 \end{array} \right| = 0 \tag{5.11}$$

Since an integral number of half wave-lengths must be completed within the molecule, we have

$$\lambda = \frac{\pi\, l}{N} \tag{5.12}$$

$$l = 0, 1, 2, \ldots, N - 1$$

By examination of the character of motion associated with each mode of vibration we see that the two vibrations of $l = 1$ appear in the Raman effect as strong lines. One of them is the symmetric deformation vibration of the chain which corresponds to the longitudinal vibration of a continuous rod referred to above and the other to the symmetric stretching vibration of the chain which remains almost constant in frequency throughout the homologous series. Using the same values of force constants as those of Equation (5.7) we can obtain the values of the deformation frequencies and the stretching frequencies, which are in good agreement with the observed values. (See Table 5.5.) It is worthy of note that the calculated frequencies of n-butane and n-pentane referred to above fall smoothly in line with the frequency values of long chain molecules beginning with hexane. This gives another justification for the assignment stated above.

TABLE 5.5

The calculated and the observed skeletal frequencies (in cm.$^{-1}$) of n-paraffin molecules in the extended form

| Number of carbon atoms | Deformation frequency | | Stretching frequency | |
|---|---|---|---|---|
| | obs. | calc. | obs. | calc. |
| 4 | 425 | 432 | 837 | 835 |
| 5 | 406 | 402 | 869 | 851 |
| 6 | 373 | 369 | 898 | 893 |
| 7 | 311 | 325 | 905 | 890 |
| 8 | 283 | 289 | 899 | 888 |
| 9 | 249 | 259 | 888 | 888 |
| 10 | 231 | 235 | 886 | 887 |
| 12 | 194 | 198 | 892 | 887 |
| 16 | 150 | 150 | 888 | 887 |

From the values of the force constants of Equation (5.7) we can estimate the energy necessary to change the C–C distance by 0.1Å as 2.9 kcal./mole and the energy necessary to change the C–C–C bond angle by 10° as 0.7 kcal./mole.[18] We have not included the torsional motion in the calculation of normal vibrations referred to above. However, from the height of the potential barrier to internal rotation it will be seen that the energy necessary to twist one part of the molecule against the other is small, so that the deformation of the paraffin molecule can take place quite easily by the internal rotation about the C–C axis.

### Summary of Chapter V

As there are three potential minima in one internal rotation about each C–C single bond, many different configurations are conceivable for a paraffin molecule. The Raman lines arising from some of these configurations can be observed in the liquid state, but the lines observed in the solid state are all assigned to the vibrations of fully extended form. This form is one of the stable configurations in the liquid state, so long as the number of carbon atoms is not large. However, in cetane with sixteen carbon atoms the Raman lines arising from the extended form is no more detectable in the liquid state. This may have to do with the theory of segments in which we consider that the long hydrocarbon chains do not flow as a single unit, but tend to flow as segments. The fact that paraffin molecules exist only in the extended form in the solid state and in various forms in the liquid state seems to be related to the alternation and convergence of melting points of $n$-paraffin series.

In the solid state we observe only one Raman line in the lower frequency region for each member of $n$-paraffin series. This can be shown to correspond to a deformation vibration of the extended chain, or a longitudinal motion of a rod, if the carbon chain is approximated by a continuous rod. This Raman line is very characteristic of the length of the chain and we can tell the number of carbon atoms of the chain molecule from the frequency value of the line.

The normal vibrations of the carbon skeleton have been calculated with reasonable values of force constants. From these values we can estimate the energy necessary to change the interatomic distance and the bond angle in the chain.

# CHAPTER VI

# Polypeptides and Related Compounds

As stated in the beginning of this book, the structure of molecules was originally studied by chemical methods of investigation including the study of the nature of chemical reactions in which substances take part. This means that at least some of the chemical reactions are characterized by the spatial configurations of molecules involved in the reactions, for example the characteristic reactions of various isomeric substituted benzenes can reasonably be explained on the basis of the hexagonal structure of the benzene nucleus.

Similarly it has been recognized recently that some biological reactions are closely related to the molecular configurations of the proteins which are the most important substances found in living organisms, e. g. the characteristic manner of the folding of polypeptide chains seems to be very important for an understanding of serological reactions. Therefore, structural studies of proteins, especially of the polypeptide chain, are very interesting not only from the view point of structural chemistry but also from that of biology.

The structural studies of proteins may be divided into two categories: (1) investigations made directly on proteins and (2) determinations of molecular and crystal structures of amino acids, amides and other simple substances closely related to proteins. Recent advances in both the experimental and theoretical sides of structural chemistry encourage the hope that the time is not far distant when we shall know the detailed structure of many protein molecules by the application of these two complementary programs.

As to the latter program, which constitutes the subject matter of the present chapter, interatomic distances and bond angles have been determined in many such substances through the work of Pauling, Corey et al. but no laboratory except ours seems to have been interested in the problem of the internal rotation of the polypeptide chain. This is an important problem in determining the structure of proteins, since without a knowledge of the internal rotation potential it would not be possible to consider any definite configuration of a polypeptide chain which would reasonably explain the crystallization of some proteins and the biological specificity. If, for example, one assumes free rotation, as many authors have done, there would be an infinite number of different configurations of polypeptide chains which would not make up a single crystal, nor would they show any specificity.

The polypeptide chain contains three kinds of single bonds:

$$-NH-CO-CHR-NH-$$
$$\ \ \ \ I\ \ \ \ \ \ II\ \ \ \ \ \ III$$

Of these three the internal rotation about the N–C axis has not been discussed at all in the foregoing chapters. Let us begin with the discussion of this problem.

### 20. Molecules with One Peptide Bond

The simplest molecule which contains a peptide bond and which may be considered as a fragment of a polypeptide chain is N-methylacetamide $CH_3CONHCH_3$. The structure determination of this molecule would, therefore, be the first problem which we have to attack in relation to the configuration of a polypeptide chain. The Raman effect, infrared absorption, dielectric constant and ultraviolet absorption of this substance have been measured by Mizushima, et al.[1] and based on these experimental results we shall discuss in the following the molecular configuration of this substance.

If now the peptide bond has pure single bond character corresponding to the valence formula $CH_3-C-NH-CH_3$, this substance would be expected
$$\overset{\|}{O}$$
to show an ultraviolet absorption similar to that of acetone, $(CH_3)_2CO$. However in an aqueous solution of this substance the absorption maximum has been found at a wave length much shorter than that of acetone and we are led from a consideration of the electronic structure of this molecule to conclude that there is a resonance,

$$CH_3C\text{-}NHCH_3 \longleftrightarrow CH_3C = N^+HCH_3 \qquad (6.1)$$
$$\ \ \ \overset{\|}{O} \qquad\qquad\qquad\quad \overset{|}{O^-}$$

for the normal state of this molecule. This conclusion is supported from the fact that, in a strongly acidic solution of this compound, an absorption maximum is observed at nearly the same wave-length as that of acetone. This absorption can reasonably be assigned to N-methylacetamidonium ion

---

1.  S. Mizushima, T. Shimanouchi, S. Nagakura, K. Kuratani, M. Tsuboi, H. Baba and O. Fujioka, J. Am. Chem. Soc., **72**, 3490 (1950).

whose normal state can approximately be represented by a single valence bond structure, $CH_3C-N^+H_2CH_3$ and which, therefore, would show similar

$$\overset{\|}{O}$$

behavior in CO absorption to acetone.

We have good reason to believe that the absorption intensity of N-methylacetamidonium ion would be of the same order of magnitude as that of acetone. Hence, if we compare the intensity of absorption of the acidic solution of N-methylacetamide with that of acetone, we can obtain the dissociation constant $K$ of the reaction, $CH_3CON^+H_2CH_3 \rightleftharpoons CH_3CONHCH_3 + H^+$,

$$K = C_2 C_3 / C_1$$

where $C_1$, $C_2$ and $C_3$ denote, respectively, the concentrations of $CH_3CON^+$ $H_2CH_3$, $CH_3CONHCH_3$ and $H^+$.

Following Pauling[2] let us attribute the difference in the dissociation constant $K$ between amine and amide to the resonance energy $E_r$ of the latter. We have then

$$E_r = R\,T \ln \frac{K[CH_3CON^+H_2CH_3]}{K[(CH_3)_2N^+H_2]} = 16\,\text{kcal./mole}$$

From this value of the resonance energy we see that the peptide N–C bond has considerable double-bond character and accordingly among the internal rotation states about this bond as axis only the *trans* and the *cis* positions are favored. Furthermore, we see that the hydrogen atom of the NH group tends to form a strong hydrogen bond and, by assuming similar resonance for the polypeptide chain, we may explain the strong tendency of a polypeptide chain to form *intramolecular* and *intermolecular* hydrogen bonds. (See next section.)

The skeleton of the N-methylacetamide molecule consists of five atoms (3 C, O and N atoms) and accordingly the skeletal frequencies number $3 \times 5 - 6 = 9$. According to the result of the Raman and infrared measurements shown in Table 6.1, almost the expected number of skeletal frequencies has been observed and, therefore, we can conclude that there is only one configuration for this molecule. The fact that the Raman spectrum of this substance remains essentially the same at higher temperatures (90–95°) and in aqueous solutions supports this view. Now this configuration must be either the *trans* or the *cis*. The calculation of the normal frequencies

---

2. L. Pauling, The Nature of Chemical Bond, Cornell University Press, p. 207 (1940).

for these two configurations leads to the result that the observed frequencie
are in good agreement with those calculated for the *trans* configuratio

$$\underset{O}{\overset{CH_3}{\diagdown}}C-N\underset{CH_3}{\overset{H}{\diagup}} \quad *$$

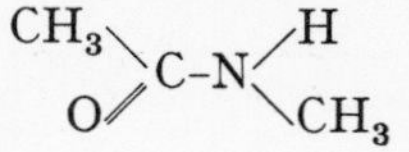

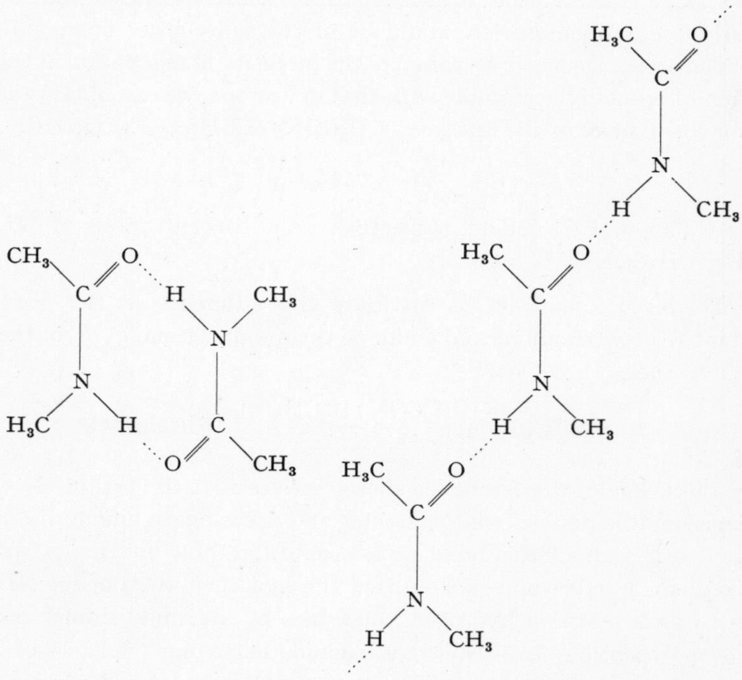

*a*         *b*

Fig. 6.1. Associated molecules of N-methylacetamide.

The results of the dipole measurements also support the view that almos
all the molecules are in the *trans* configuration. The concentration-dependenc
of the molecular polarization observed in carbon tetrachloride solution
indicates that this molecule is associated considerably in such solutions

---

 * One might be interested in the detection of Raman lines corresponding t
N-methylacetamidonium ion in an acidic solutions referred to above. Howeve
the calculation based on the ultraviolet absorption data shows that for a solutio
of p H $= 0$, the number of molecules in the ionic form amounts only to on
percent of that in the normal form. It is, therefore, no wonder that we coul
not observe the Raman lines of the ionic form in the acidic solution.

| Raman effect liquid | Infrared absorption | | |
|---|---|---|---|
| | vapor | liquid | solid (−60 °C) |
| 206(1 b) | | | |
| 289(2 b) | | | |
| 435(4) | | | |
| 600(0) | | 600 (m, b) | 615 (m) |
| 628(6) | | 628 (m) | 629 (m) |
| | | 725 (m, b) | — |
| | | — | 800 (m) |
| | 810 (w) | | |
| 879(7) | | 876 (w) | 891 (m) |
| | 959 (w) | | |
| 992(2) | | 989 (m) | 1001 (m) |
| 1042(1) | | 1039 (m) | 1050 (w) |
| 1096(1) | | 1095 (m) | 1097 (w) |
| 1164(5) | | 1158 (s) | 1161 (s) |
| | 1247 1261 (s) | | |
| 1298(5) | | 1298 (s) | 1312 (s) |
| 1374(3) | 1371(m) | 1374 (s) | |
| | 1382 (m) | | 1385 (s) |
| 1415(3) | | 1413 (s) | 1415 (m) |
| 1447(2 b) | | 1442 (m) | 1441 (m) |
| | | | 1457 (sh) |
| | 1487 (vs) | | |
| | 1531 (sh) | 1531 (sh) | 1530 (sh) |
| | | 1564 (vs) | 1583 (vs) |
| 1657(5 b) | | 1652 (vs) | 1635 (s) |
| | 1717 (vs) | | |
| | 1730 (sh) | | |
| 2724(0 b) | | 2720 (w) | 2720 (w) |
| | | 2890 (sh) | 2910 (s) |
| 2934(8) | 2940 (s) | 2940 (s) | |
| 2994(3 b) | | | 2985 (s) |
| | | 3110 (vs) | 3105 (vs) |
| 3300(2 bb) | | 3300 (vs) | 3250 (vs). |
| | 3500 (m) | | |

If this molecule were in the *cis* form it would be associated to form a dimer just as in the case of a carboxylic acid (Fig. 6.1 *a*) and the apparent moment would become larger with decreasing concentration. This is not compatible

with the results of dipole measurements in which the apparent moment ha
been found to decrease from 6.6 D to 4.8 D as the weight fraction of solut
was lowered from 0.006 to 0.0006. If, however, the single molecule has th
*trans* configuration as proposed by the Raman and infrared measurements
the associated molecule will have a chain configuration which has a momen
value larger than that of a single molecule, and which, therefore, woul
account for the concentration-dependence of the apparent moment quit
well (Fig. 6.1 *b*).

This conclusion is supported by the dipole measurements made for sucl
a molecule as $\delta$-valerolactam which has a ring structure and whose C=C
and N–H groups are nearly in the *cis* position to each other. (See Fig. 6.2.

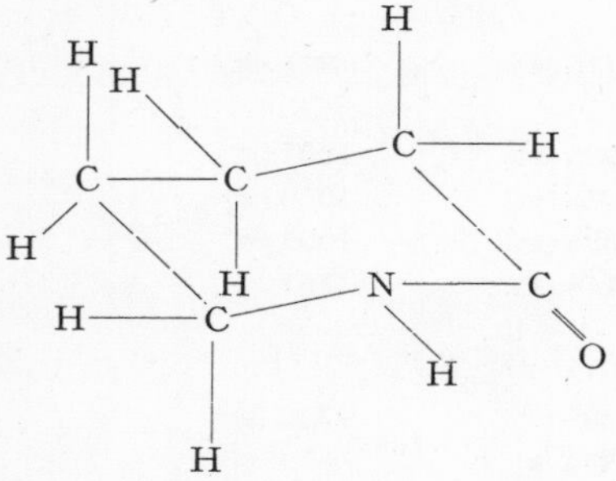

Fig. 6.2. Molecular configuration of
$\delta$-valerolactam.

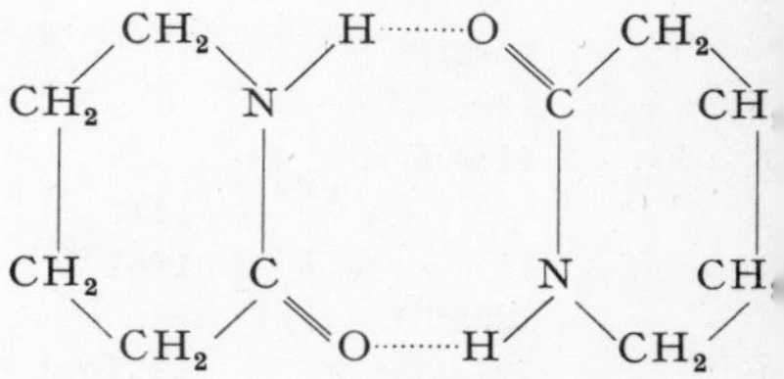

Fig. 6.3.
Associated molecules of $\delta$-valerolactam.

In carbon tetrachloride solution at 20° C the value of the apparent moment
has been found to be 2.33, 2.24 and 2.09 $D$ for the weight fractions 0.0052
0.0083 and 0.0138, respectively. The moment value is much smaller than
that of N-methylacetamide and the apparent value increases with decreasing
concentration. This would be naturally explained, if we assume a ring dimer
shown in Fig. 6.3 which is characteristic of the *cis* configuration of the single
molecule.

The near infrared absorption of this substance in the $3\mu$ and the $1.5\mu$
regions has also been measured by the same authors.[1] They observed absorption peaks at 3500 cm.$^{-1}$ ($2.86\mu$) in the gaseous state and at 3470 cm.$^{-1}$
($2.88\mu$) in dilute carbon tetrachloride solutions and assigned these peaks
to the N–H fundamental vibrations in the free state. The corresponding
first overtones was observed at 6760 cm.$^{-1}$ ($1.48\mu$). At higher concentrations,
however, the 3470 cm.$^{-1}$ absorption becomes weaker and at the same time

new absorptions are observed at $3370 \sim 3300$ cm.$^{-1}$ and at $3110$ cm.$^{-1}$ $(3.22\mu)$, both arising from the associated molecules. The continuous range of wave-length of the former band may be due to associated molecules of different chain lengths. The appearance of the two absorption peaks for the associated N–H vibrations has not yet been explained satisfactorily. According to Oshida and Oshika this might be due to the presence of two stable positions of the proton of N–H...O separated by a potential barrier of suitable height.[3]

In the liquid state and in the crystal the $3470$ cm.$^{-1}$ band or the band due to N–H vibration of the single molecule disappears completely and only the bands arising from the associated molecules are observed. (See Table 6.2.)

As shown in Table 6.1 the CO absorption (the CO stretching frequency) is observed at $1652$ cm.$^{-1}$ in the liquid state. In carbon tetrachloride solution this absorption is shifted toward higher frequency (up to about $1700$ cm.$^{-1}$), as the concentration is lowered. This fact confirms the view that the association of N-methylacetamide is effected through N–H...O hydrogen bond. The frequency values of the CO absorption ($1652 - 1700$ cm.$^{-1}$) are lower than those of ketones (about $1740$ cm.$^{-1}$). This is attributable to the resonance contribution of the form $\overset{}{\underset{O-}{\diagdown}}C{=}N^{+}\overset{H}{\diagup}$ to the electronic structure of this group, as has been shown from the ultraviolet absorption studies.

We have already stated that $\delta$-valerolactam is associated in a manner quite different from that of N-methylacetamide. This is also shown from the different behavior of the N–H association band. While this band undergoes a continuous change of frequency in the case of N-methylacetamide, no such change is observed in the case of $\delta$-valerolactam. This should be so, if this substance is associated to form a double molecule of definite configuration. Actually $\delta$-valerolactam in carbon tetrachloride solution shows the free

TABLE 6.2

Frequencies (cm.$^{-1}$) of absorption peaks of N-methylacetamide in the $3\mu$ region[1]

| State | | Temp. (°C) | Free N–H | Associated N–H | |
|---|---|---|---|---|---|
| | Gas | 210 | 3500 | — | |
| CCl$_4$ soln. | 0.0014 mole/l. | 15 | 3470 | — | |
| | 0.03–0.08 mole/l. | room temp. | 3470 | 3370–3300, | 3110 |
| | Liquid | 30 | — | 3300, | 3110 |
| | Crystal | 10 | — | 3300, | 3110 |

3. I. Oshida and Y. Oshika  Busseiron-Kenkyu, 46, 95 (1952).

N–H band at 3420 cm.$^{-1}$ ($2.92\mu$) and the association bands at 3220 cm.$^{-}$ ($3.11\mu$) and at 3090 cm.$^{-1}$ ($3.24\mu$), of which the latter two bands decrease in intensity on dilution, but do not undergo any frequency change. Furthermore, the association bands of this substance can be observed at a dilution at which no association bands of N-methylacetamide can be detected. This may be taken as further evidence that the molecule of $\delta$-valerolactam is associated through double hydrogen bonds, while that of N-methylacetamide is associated through a single hydrogen bond.*

It is worthy of note that the main chain of polypeptide shows near infrared absorption quite similar to that of the associated N-methylacetamide but different from that of $\delta$-valerolactam. Accordingly, it is probable that the peptide bond of the polypeptide chain has the planar *trans* configuration. We shall show in the following section that the peptide bond also keeps the planar *trans* form in molecules with two peptide bonds, although these molecules can exist in different configurations.

### 21. Molecules with Two Peptide Bonds

We now pass on to the explanation of the structure of molecules of the type $R_1CONHCHRCONHR_2$ containing two peptide bonds and shall first discuss the near infrared absorption bands as we have done for molecules containing one peptide bond. In Table 6.3 are shown the frequencies $\nu$ of absorption peaks and the molar absorption coefficients of acetylglycine N-methylamide $CH_3CONHCH_2CONHCH_3$, acetylglycine anilide $CH_3CONH$ $\cdot CH_2CONHC_6H_5$, acetyl-DL-leucine N-methylamide $CH_3CONHCHCONHCH_3$

$$\overset{|}{C}H_2CH(CH_3)_2$$

and acetyl-DL-leucine anilide $CH_3CONHCHCONHC_6H_5$ measured by

$$\overset{|}{C}H_2CH(CH_3)_2$$

Mizushima et al.[4]

---

* For the near infrared spectra of compounds containing the peptide bond, see also the following references:

A. M. Buswell, J. R. Downing and W. H. Rodebush, J. Am. Chem. Soc., **62**, 2759 (1940).

J. Lecomte and R. Freymann, Bull. soc. chim. **8**, 601 (1941).

R. E. Richards and H. W. Thompson, J. Chem. Soc., **1947**, 1248.

S. E. Darmon and G. B. B. M. Sutherland, Nature **164**, 440 (1949); J. Am. Chem. Soc., **69**, 2074 (1947).

S. Mizushima, T. Shimanouchi and M. Tsuboi, Nature, **166**, 406 (1950).

---

4. S. Mizushima, T. Shimanouchi, M. Tsuboi, T. Sugita, E. Kato and E. Kondo, J. Am. Chem. Soc., **73**, 1330 (1951).

From the discussions given in the preceding section it will be evident that he absorption peak at 3450 cm.$^{-1}$ (2.9$\mu$) arises from the NH group in the ree state and the one at 3330 cm.$^{-1}$ (3.0$\mu$) arises from the group involved a the hydrogen bonding. Since these two absorption peaks were exhibited •y all the compounds shown in Table 6.3, at least some of their molecules hould contain both the free and the hydrogen-bonded NH groups. The act that $\varkappa$ of the 3.0$\mu$ band as well as $\varkappa$ of the 2.9$\mu$ band are practically ndependent of the concentration at a fixed temperature (Table 6.3) would

TABLE 6.3

requencies $\nu$ (cm.$^{-1}$) of absorption peaks and molar absorption coefficients $\varkappa$ of the $N$-H bands of $CH_3CONHCHRCONHCH_3$ and $CH_3CONHCHRCONHC_6H_5$ in dilute $CCl_4$ solutions

| Substance | Temp. (°C) | Concn. (mole/l.) | Free N–H | | Hydrogen-bonded N–H | |
|---|---|---|---|---|---|---|
| | | | $\nu$ | $\varkappa$ | $\nu$ | $\varkappa$ |
| Acetylglycine N-methyl-amide | 60 | 0.000165 | 3450 (2.90$\mu$) | 90 | 3360 (2.98$\mu$) | 44 |
| | 60 | 0.000315 | 3450 (2.90$\mu$) | 94 | 3360 (2.98$\mu$) | 44 |
| | 60 | 0.000525 | 3450 (2.90$\mu$) | 94 | 3360 (2.98$\mu$) | 40 |
| | 60 | 0.000765 | 3450 (2.90$\mu$) | 89 | 3360 (2.98$\mu$) | 45 |
| | 20 | 0.000165 | 3450 (2.90$\mu$) | 85 | 3360 (2.98$\mu$) | 82 |
| | 20 | 0.000315 | 3450 (2.90$\mu$) | 87 | 3360 (2.98$\mu$) | 85 |
| Acetylglycine anilide | 60 | 0.00014 | 3450 (2.90$\mu$) | 99 | 3330 (3.00$\mu$) | 72 |
| | 60 | 0.000285 | 3450 (2.90$\mu$) | 100 | 3330 (3.00$\mu$) | 77 |
| | 60 | 0.00045 | 3450 (2.90$\mu$) | 95 | 3330 (3.00$\mu$) | 68 |
| | 20 | 0.00014 | 3450 (2.90$\mu$) | 95 | 3330 (3.00$\mu$) | 109 |
| | 20 | 0.000285 | 3450 (2.90$\mu$) | 97 | 3330 (3.00$\mu$) | 126 |
| Acetyl-DL-leucine N-methylamide | 60 | 0.000555 | 3450 (2.90$\mu$) | 82 | 3360 (2.98$\mu$) | 44 |
| | 20 | 0.00032 | 3450 (2.90$\mu$) | 75 | 3360 (2.98$\mu$) | 64 |
| Acetyl-DL-leucine anilide | 60 | 0.000155 | 3450 (2.90$\mu$) | 61 | 3330 (3.00$\mu$) | 36 |
| | 20 | 0.00016 | 3420 (2.92$\mu$) | 52 | 3310 (3.02$\mu$) | 74 |

ndicate that the hydrogen bond is *intramolecular*. This conclusion is supported by the fact that N-methylacetamide, acetanilide, etc., which form only the *intermolecular* hydrogen bond show practically no absorption due to the hydrogen-bonded NH groups in such high dilutions. As to the molecular configuration with internal hydrogen bond, we can consider one such as:

$$\begin{array}{c}
\text{R} \\
| \\
\text{H} \diagdown \quad \text{CH} \diagdown \\
\text{N} \diagdown \quad \quad \text{C.} \diagup \text{O} \\
| \quad \quad \quad \vdots \\
\diagup \text{C} \diagdown \quad \quad \text{HN} \diagdown \\
\text{R}_1 \quad \quad \text{O} \quad \quad \quad \text{R}_2
\end{array}$$

This is quite possible from the viewpoint of internal rotation potential, since all the atoms or groups are in or near the potential minima of internal rotation about the single bonds contained in the molecule. (The –CO–NH– peptide bond is in the planar *trans* form.) Another possible configuration from the same point of view is the extended form:

$$\begin{array}{c}
\text{O} \quad \quad \text{R} \quad \quad \text{H} \\
\| \quad \quad | \quad \quad | \\
\text{C} \diagup \quad \text{CH} \diagup \quad \text{N} \\
\text{R}_1 \diagup \quad \text{N} \diagdown \quad \text{C} \diagup \quad \text{R}_2 \\
| \quad \quad \| \\
\text{H} \quad \quad \text{O}
\end{array}$$

This, however, forms only *intermolecular* hydrogen bonds, but not *intramolecular* hydrogen bonds. The experimental results so far stated do not provide positive evidence for the presence of this form, but they show definitely the presence of the folded or the bent form in dilute carbon tetrachloride solutions.

If only the latter form is present in carbon tetrachloride solutions, the intensity ratio of the $2.9\mu$ band (free N–H) to the $3.0\mu$ band (bonded N–H) should be independent of temperature. However, we actually observe a conspicuous change of the intensity ratio with temperature, as can be readily seen from the $\varkappa$-values of the $2.9\mu$ and the $3.0\mu$ bands shown in Table 6.3. Hence it follows that the extended form should be present in carbon tetrachloride solutions in addition to the folded form and the equilibrium ratio of these two forms changes greatly with temperature. Of interest in this connection is the observation of the near infrared spectra of acetylproline N-methylamide.[5] It was found that only the $3.0\mu$ band appears in the $3\mu$ region in the measurement covering a wide range of concentration. Accordingly, we have only the hydrogen-bonded N–H group in the molecule of acetylproline N-methylamide even in a very dilute carbon tetrachloride solution. The only possible explanation for this experimental result is to

5. S. Mizushima, T. Shimanouchi, M. Tsuboi, T. Sugita, K. Kurosaki, N. Mataga and R. Souda, J. Am. Chem. Soc., **74**, 4639 (1952).

assume that practically all the molecules of this substance take the folded form:

with no NH group in the free state. This conclusion is compatible with the fact that acetylproline N-methylamide is much more soluble in a nonpolar solvents than acetylglycine N-methylamide or acetyl alanine N-methylamide, since in the folded configuration all the hydrogen bonds are *intramolecular*.[6] The value of the dipole moment observed for this substance (3.3 D) is also in agreement with the theoretical value calculated for the folded form.[6]

For the other acetylaminoacid N-methylamides stated above we have an equilibrium between the extended and folded forms, which can be studied by the observation of the temperature-dependence of the infrared absorption intensity. Let the energy and the entropy of the extended form be larger by $\Delta E$ and $\Delta S$, respectively, than those of the folded form and let $\varkappa'$ and $\varkappa$ be the molar absorption coefficient of the $3.0\mu$ band for the pure folded form and that for the mixture of the folded and extended forms, respectively. Then we have

$$\varkappa = \varkappa' \Big/ \left( 1 + e^{\frac{\Delta S}{R} - \frac{\Delta E}{RT}} \right), \tag{6.2}$$

since the $3.0\mu$ band arises only from the folded form in a very dilute solution. From the data of the energy difference between the rotational isomers so far collected, we can consider that the value of $\Delta E$ is less than 6 kcal./mole. Introducing the values of $\Delta E$ in this range into Eq. (6.2), we can show the entropy difference between the extended and folded forms to be greater than 15 e. u. This is a fairly large value which is at least partly explained by considering that the freedom of the internal rotation of the extended form is much greater than that of the folded form.

---

6. S. Mizushima, T. Shimanouchi, K. Kuratani, M. Tsuboi, T. Sugita, I. Nakagawa and K. Kurosaki, Nature, **169**, 1058 (1952).

The conclusions drawn above for the structure of the molecules with two peptide bonds are useful in constructing a stable model of polypeptide chain. The tendency to take on an extended form or a folded form for acetylaminoacid N-methylamide may be closely related to the tendency of taking one of these forms for the corresponding aminoacid residue in the polypeptide chain. Accordingly, the proline residue would show a strong tendency to take the folded form*.

The entropy difference between the extended and the folded forms of acetylaminoacid N-methylamides may be related to the entropy change of a polypeptide chain on extension of the folded form. If, therefore, the denaturation of proteins is associated with the unfolding of the polypeptide chain, we may explain the large entropy change on denaturation by the value of $\Delta S$ stated above.

Returning to the problem of the absorption intensity of the N–H band, the molecular absorption coefficient $\varkappa$ of N-methylacetamide referred to the same standard as that of Table 6.3 was found to be 20. If, therefore, the free N–H groups of the compounds shown in the same table have the same bond moment as that of N-methylacetamide, the values of $\varkappa$ for the 2.9$\mu$ band of these compounds should not be much different from 20, in so far as their molecules are in the folded form. And even if all the molecules were in the extended form with twice as many free N–H groups as in the case of the folded form, the value of $\varkappa$ should not exceed 40. Therefore, the observed values of $\varkappa$ amounting to 70 $\sim$ 100 (see Table 6.3) indicate that the bond moment of the free NH group of the compounds with two peptide bonds is greater than those with one peptide bond. This would be explained by considering that the contribution of the resonance structure:

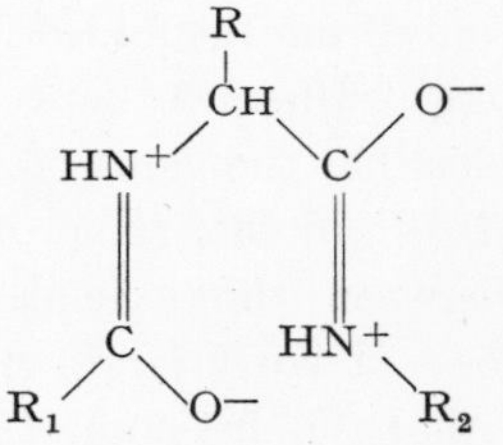

$$
\begin{array}{c}
R \\
| \\
CH \\
HN^+ \diagup \quad \diagup C \diagdown \quad O^- \\
\| \qquad \qquad \| \\
C \qquad \quad HN^+ \\
R_1 \diagup \quad \diagdown O^- \qquad \diagdown R_2
\end{array}
$$

---

* It should be realized that in a polar solution, in which the hydrogen bonding is reduced, the proline residue would partly take a configuration or configurations other than the folded form.

for the folded form is made larger by the *intramolecular* hydrogen bonding
(see below). Therefore, the peptide bond

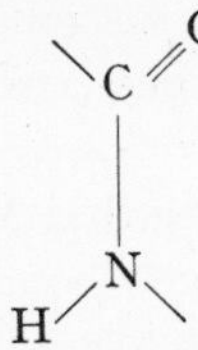

definitely has the planar configuration and is in the *trans* form, which is
compatible with the fact that in this case we have also found the bonded
N–H band at $3.0\mu$ and not at $3.1\mu$ (see the preceding section). This conclusion

will also have an important relation to
the structure of the peptide bonds
contained in the polypeptide chain.

Now let us discuss the more concentrated solutions of acetylleucine N-
methylamide in order to study the
association of the molecules with two
peptide bonds. As the concentration is
raised, the intensity of the $2.9\mu$ band
(the free N–H band) becomes weaker
until this band becomes hardly detectable, while the intensity of the $3.0\mu$
band (the bonded NH band) becomes
stronger and the frequency of the peak
of this band is shifted gradually from
$3360$ cm.$^{-1}$ $(2.98\mu)$ to $3300$ cm.$^{-1}$ $(3.03\mu)$.
At the same time the $3110$ cm.$^{-1}$ $(3.22\mu)$
band begins to appear. All these
phenomena are quite the same as have
been found for N-methylacetamides (see
preceding section) for which we have
deduced a chain type of association.
However, the molecular association of
acetylleucine N-methylamide in carbon

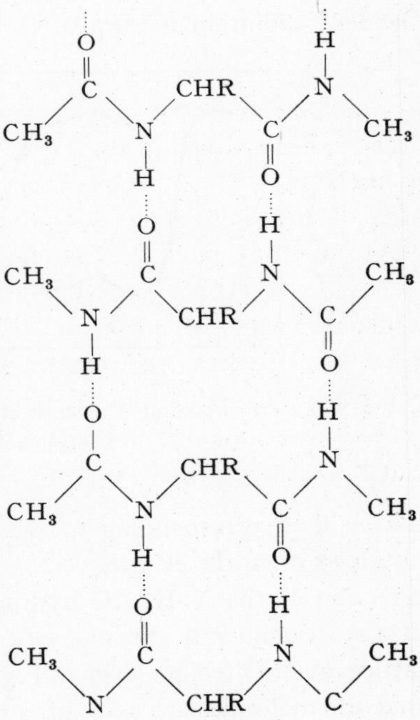

Fig. 6.4. Associated molecules of
acetylleucine N-methylamide.

tetrachloride solutions is much stronger than that of N-methylacetamides.
For example, there is a considerable number of associated molecules at the
concentration of 0.001 mole/l., at which almost all the molecules of
N-methylacetamide are found to exist in the single state. This will be

explained as due to the fact that the extended form of acetylleucine N-methylamide can form double hydrogen bonds as shown in Fig. 6.4 and accordingly, this substance will show stronger association as compared with N-methylacetamide whose molecules associate through single hydrogen bond. (See Fig. 6.1.)

In this connection it is interesting that Mizushima et al.[7] found molecular aggregation through double hydrogen bonds in the crystal of acetylleucine N-methylamide. As shown in Fig. 6.5 only the bonded NH band is observed at 3290 cm.$^{-1}$ ($3.04\mu$) and 3110 cm.$^{-1}$ ($3.22\mu$) and only the bonded CO band at 1616 cm.$^{-1}$ ($6.19\mu$) in the crystalline state. These bands show pronounced dichroism. The full line of Fig. 6.5 corresponding to the strong absorption of the NH and CO bands is obtained when the electric vector of the polarized incident radiation is parallel to the long direction of the crystal and the

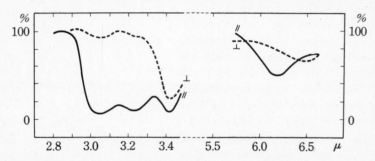

Fig. 6.5. Transmission of polarized infrared radiation in the oriented crystal of acetylleucine N-methylamide.

dotted line corresponding to the very weak, or almost zero, absorption is obtained when the electric vector is perpendicular to it. This means that the direction of the N–H...O hydrogen bond is parallel to the long direction of the crystal and the molecules of acetylleucine N-methylamide in the lattice are oriented as shown in Fig. 6.4. Therefore, it seems natural to assume that the molecules are associated in a similar manner in carbon tetrachloride solutions: in other words the molecules in the extended form are associated through double hydrogen bonds. Thus we can understand the difference in the strength of association between acetylleucine N-methylamide and N-methylacetamide.

7. S. Mizushima, T. Shimanouchi, M. Tsuboi, K. Kuratani, T. Sugita, N. Mataga and R. Souda, J. Am. Chem. Soc., 75, 1863 (1953).

In the carbon tetrachloride solutions of acetyl proline N-methylamide the molar absorption coefficient as well as the frequency of the $3.0\mu$ band has been found to be almost independent of concentration, although this has been varied from 0.0005 to 0.05 mole/l. This means that even at a concentration as high as 0.05 mole/l. almost all the molecules of acetylproline N-methylamide are non-associated, due very probably to the fact that the molecule exists only in the folded form which can form no *intermolecular* hydrogen bond.[5]

As explained in detail in this section molecules of the type $CH_3CONH$ $\cdot CHRCONHCH_3$ exist either in the extended form or in the folded form. From the analogy in chemical constitution between this type of molecule and the structural unit of a polypeptide chain, we may consider that a polypeptide chain also contains such extended and folded forms. We shall discuss this problem in detail in Section 23, based on the experimental data obtained for proteins.*

Before concluding the explanation of the structure of molecules with peptide bonds, it would be appropriate to discuss the strength of hydrogen bonds to some extent, because this plays an important part in the problem of the configuration of polypeptide chain as well as in several problems of simple molecules which we have been discussing.

When a group X–H is involved in hydrogen bonding (X–H...Y), the X–H absorption is shifted to lower frequencies, as in the case of the N–H...O system of N-methylacetamide, acetylglycine N-methylamide. etc. An estimation of the strength of hydrogen bonds can be made from the amount of the shifts in frequency as has been pointed out by several investigators. [8, 9]

---

* In this connection it is interesting that a configuration:

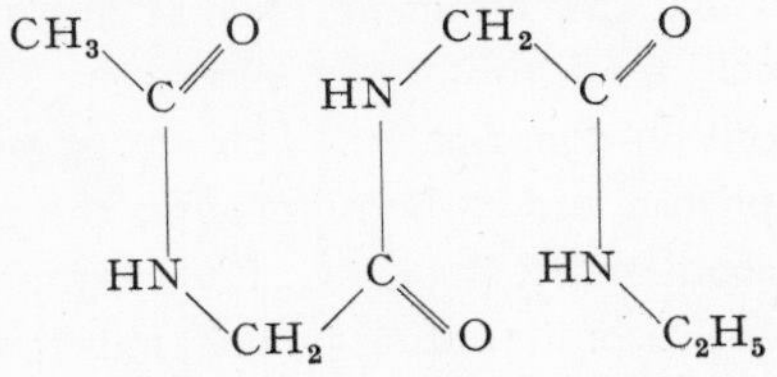

was proposed for acetylglycylglycine N-ethylamide by H. Lenormant and J. Chouteau, J. de Physiologie, **43**, 778 (1951).

---

8.   J. J. Fox and A. E. Martin, Proc. Roy. Soc., (A) **162**, 419 (1937).
9.   R. M. Badger and S. H. Bauer, J. Chem. Phys., **5**, 839 (1937).

For example, Gordy[10] has measured the strength of hydrogen bonds formed between deuterated methyl alcohol and many organic liquids from the magnitude of the shifts which these liquids produce in the OD frequency of $CH_3OD$.

Useful data concerning this problem have been obtained by Tsuboi,[11] who measured the X–H frequencies in ternary solutions which consist of large quantity of nonpolar solvent such as carbon tetrachloride and small quantities of two polar solutes. One of these solutes is a proton-donor with the X–H group and the other is a proton-acceptor with the Y atom, so that complex molecules with the hydrogen bond X–H...Y are formed in an indifferent solvent. On the basis of such data we can reasonably rate different substances according to their proton-donating power or proton-accepting power.

Tsuboi[11] measured the frequencies and the absorption intensity of the free and the bonded X–H bands at different temperatures and at different concentrations. From the results he calculated the energy of the hydrogen bond X–H...Y. By comparing the energy values obtained for different hydrogen bonds in the same medium, the following general conclusion can be derived. The proton-donating power of the X–H group is determined by the tendency of the X atom to attract the electron of hydrogen to make the proton bare, and the proton-accepting power of the Y atom is determined by the tendency of this atom to attract electrons from the adjacent atoms or groups to increase its effective negative charge. Evidently, a strong hydrogen bond, X–H...Y, is formed, when the X–H group has the strong proton-donating power and the Y atom has the strong proton-accepting power.

The amino groups of alkylamines have weak proton-donating power, but have strong proton-accepting power. Their proton-donating power becomes stronger, when the adjacent groups are phenyl, carbonyl and sulfonyl groups, which tend to attract electrons. The hydroxyl group behaves as a strong proton-donor as well as a strong proton acceptor. If the adjacent groups are phenyl and carbonyl groups, the proton-donating power of hydroxyl group becomes still stronger. For example, the hydroxyl group of phenol has stronger proton-donating power than that of aliphatic alcohol.

---

10. W. Gordy, J. Chem. Phys., **7**, 93 (1939); W. Gordy and S. C. Stanford, **8**, 170 (1940), **9**, 204, 215 (1941).

11. M. Tsuboi, Bull. Chem. Soc. Japan, **25**, 60 (1952); J. Chem. Soc. Japan, **72**, 146 (1951).

However, as the oxygen atom of phenol is weaker in proton-accepting power than that of aliphatic alcohol, the association of phenol is not stronger than that of aliphatic alcohols. In the following table various groups are rated according to their proton-donating or proton-accepting power:

*Proton-donating power*

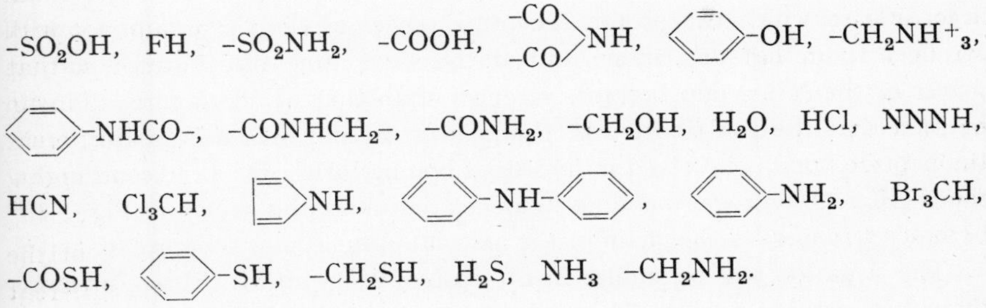

$-SO_2OH$, FH, $-SO_2NH_2$, $-COOH$, $\begin{smallmatrix}-CO\\-CO\end{smallmatrix}\!>\!NH$, ⟨⟩$-OH$, $-CH_2NH^+{}_3$,

⟨⟩$-NHCO-$, $-CONHCH_2-$, $-CONH_2$, $-CH_2OH$, $H_2O$, HCl, NNNH,

HCN, $Cl_3CH$, ⟨⟩$NH$, ⟨⟩$-NH-$⟨⟩, ⟨⟩$-NH_2$, $Br_3CH$,

$-COSH$, ⟨⟩$-SH$, $-CH_2SH$, $H_2S$, $NH_3$ $-CH_2NH_2$.

*Proton-accepting power*

$\begin{smallmatrix}CH_2\\ \end{smallmatrix}$ $\begin{smallmatrix}CH_2\\ \end{smallmatrix}$ $\begin{smallmatrix}CH_2\\ \end{smallmatrix}$ $CH_2$ $CH_2$ $CH_2$

$NH$, $NH_2$, $\diagdown{N}\diagup$ , $NH_3$, ⟨⟩$_N$ , $-COO^-$, $\underset{\underset{O}{\|}}{C}OH$,

⟨⟩$NH_2$, ⟨⟩$NHCH_2-$, ⟨⟩$N\!\!<\!\!\begin{smallmatrix}CH_2-\\CH_2-\end{smallmatrix}$, $R\!-\!\underset{\underset{O}{\|}}{\overset{\overset{O}{\|}}{S}}\!-\!NH_2$, $R\!-\!\underset{\underset{O}{\|}}{C}\!-\!NH_2$,

$R\!-\!\underset{\underset{O}{\|}}{C}\!-\!NH\!-\!CH_2-$, $R\!-\!\underset{\underset{O}{\|}}{C}\!-\!SH$, ⟨⟩$NH$⟨⟩, ⟨⟩$_{\underset{H}{N}}$ , $H_2O$, $-CH_2OH$,

$\begin{smallmatrix}-CH_2\\-CH_2\end{smallmatrix}\!>\!O$, $\underset{\underset{O}{\|}}{-CH}$, $\begin{smallmatrix}-CH_2\\-CH_2\end{smallmatrix}\!>\!C\!=\!O$, $-CN\!\equiv\!N$, $-NO_2$, ⟨⟩$-OH$,

⟨⟩$-OR$, NNNH, $H_2S$, $-CH_3SH$, ⟨⟩$-SH$, $\diagup\!\!>\!\!C-Cl$, $\diagup\!\!>\!\!C-Br$.

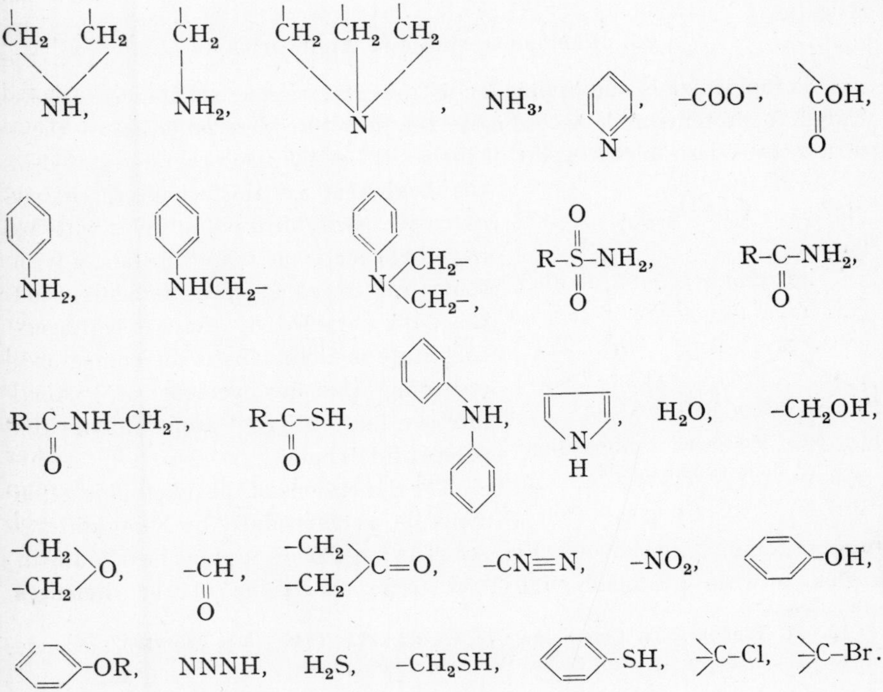

The hydrogen bond formed between N–H and O=C groups, which we have been discussing in this section and in the immediately preceding one, is much stronger than that formed between the similar groups of e. g. acetone and alkylamine. In the case of N-methylacetamide this is due to the resonance:

$$CH_3C\text{–}NHCH_3 \longleftrightarrow CH_3C=N^+HCH_3,$$
$$\overset{\|}{O} \qquad\qquad \overset{|}{O^-}$$

according to which the proton accepting power of the CO group becomes stronger than that of acetone and at the same time the proton-donating power of the NH group becomes stronger than that of alkylamine. It will be seen that the contribution of the polar structure becomes greater, once the peptide bond is involved in hydrogen bond, N–H...O=C. Accordingly, the N–H...O hydrogen bonding in associated molecules shown in Fig. 6.1 *b* becomes stronger, as the chain of the associated molecules becomes longer.[12]

The situation will be similar in the case of molecules with two, or more than two, peptide bonds. Therefore, we have good reason to believe that the *intramolecular* and the *intermolecular* hydrogen bonds of similar type will be formed fairly strongly in polypeptide chains or between the chains.

## 22. Diketopiperazine and Aminoacids

Diketopiperazine and amino acids are also simple substances closely related to proteins and, accordingly, the structure determination of these substances is desirable as a part of the second of the two programs stated in the beginning of this chapter. Fairly extensive X-ray diffraction investigations have been made on these substances with some spectroscopic measurements, and the data obtained by these experiments constitute a factual basis for conjectures regarding the arrangement of carbon, nitrogen and oxygen atoms along the polypeptide chain.

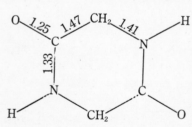

Fig. 6.6. Molecular configuration of diketopiperazine.[13]

The dimensions of the diketopiperazine molecule, as determined by X-ray diffraction by Corey,[13] are shown in Fig. 6.6. The molecule is a plane hexagon with angles between all bonds $120 \pm 3°$. The shortening of the distances

12. M. Tsuboi, Bull. Chem. Soc. Japan, **22**, 215 (1949); **25**, 385 (1952).
13. R. B. Corey, J. Am. Chem. Soc., **60**, 1598 (1938).

$CH_2$–N (1.41Å) and $CH_2$–C (1.47Å) below the normal single bond values 1.47 and 1.54Å is unusual.*

The infrared absorption spectra of diketopiperazine have been measured by Newman and Badger[14] and Shimanouchi, Kuratani and Mizushima,[15] both groups using polarized radiation. In particular the absorption in the $3\mu$ region was measured by Tsuboi[16] and Ambrose et al.[17] The experimental results are compatible with the molecular configuration shown in Fig. 6.6, in which the peptide bond is in the *cis* form as in the case of δ-valerolactam, ε-caprolactam etc., indicating that diketopiperazine belongs to a category different from that of $CH_3CONHCH_3$, $CH_3CONHCHRCONHCH_3$, etc. stated in the preceding sections.

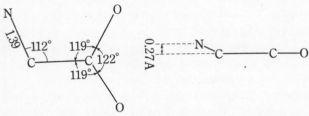

Fig. 6.7.  Skeletal configuration of glycine molecule.[18]

The configuration of the glycine molecule determined by Albrecht and Corey[18] is shown in Fig. 6.7. All distances and bond angles are normal with the exception of the distance C–N (1.39Å). The carbon and oxygen atoms are coplanar: the nitrogen atom lies slightly (0.27Å) out of the molecular plane. This is the only stable position of internal rotation about the C–C axis.

The result of crystal structure determination by the same investigators suggests strongly that the molecule in the crystal exists as a zwitterion with positively charged $NH_3$ group and negatively charged oxygen atom. The electrostatic force due to these positive and negative extremes as well as the hydrogen bonds between nitrogen and oxygen atoms firmly folds the molecules in the lattice.

---

* For the normal values of single bonds, see Section 1.

14.  R. Newman and R. M. Badger, J. Chem. Phys., 19, 1147 (1951).
15.  T. Shimanouchi, K. Kuratani and S. Mizushima, J. Chem. Phys., 19, 1479 (1951).
16.  M. Tsuboi, Bull. Chem. Soc. Japan, 22, 255 (1949).
17.  E. J. Ambrose, A. Elliott and R. B. Temple, Proc. Roy. Soc., (A) 206, 192 (1951).
18.  G. Albrecht and R. B. Corey, J. Am. Chem. Soc., 61, 1087 (1939).

The zwitterion in crystalline glycine was also suggested by the Raman measurement by Baba et al.[19] They were able to observe the characteristic frequencies of zwitterion but no Raman line which corresponds to the CO frequency of the COOH group. These investigators also measured the Raman spectra of glycine in aqueous solutions and found the structures, $NH_3^+CH_2COO^-$, $NH_3^+CH_2COOH$ and $NH_2CH_2COO^-$ in the neutral, acidic and basic solutions, respectively. Their results in solutions are in essential agreement with those of the earlier investigators, among whom Edsall and his coworkers[20] made most extensive measurements not only on glycine but also on other aminoacids and related compounds. As the first-mentioned investigators used an apparatus of high luminosity, they could observe all the lower frequency lines which are necessary for the calculation of skeletal frequencies.[19] The result of this calculation is consistent with the configuration of the glycine skeleton as proposed from the X-ray investigation referred to above. (The CCN bending frequency which is observed at 420 cm.$^{-1}$ is the most suitable one by which we can determine the configuration of glycine. In other words, if we calculate this frequency for different configurations we see that there is a conspicuous dependence of frequency upon azimuthal angle of internal rotation. We shall show in Part II that a skeletal bending frequency is in general most suitable for the determination of the azimuthal angle.)

In infrared absorption, the evidence for the zwitterion structure of glycine and other amino acids is equally strong.[21] An interesting fact first noticed by Wright[22] is that the infrared spectrum of the DL-form of an amino acid is usually different from the spectrum of either the D- or the L-form of the same acid when each is examined in the solid state. Darmon, Sutherland and Tristram[23] confirmed this observation.

19. H. Baba, A. Mukai, T. Shimanouchi and S. Mizushima, J. Chem. Soc. Japan, 70, 333 (1949).

20. J. T. Edsall, J. Chem. Phys., 4, 1 (1936); J. Phys. Chem., 41, 133 (1937); J. Chem. Phys., 5, 225, 508 (1937); Cold Spring Harbor Symposia, 6, 40 (1938); J. Am. Chem. Soc., 65, 1767 (1943). J. T. Edsall and H. Scheinberg, J. Chem. Phys., 8, 520 (1940). J. T. Edsall, J. W. Otvos and A. Rich, J. Am. Chem. Soc., 72, 474 (1950).

21. I. M. Klotz and D. M. Gruen, J. Phys. Colloid Chem., 52, 961 (1948). M. M. Davies and G. B. B. M. Sutherland, J. Chem. Phys., 6, 755 (1938). S. E. Darmon, Dissertation, Cambridge (1948).

22. N. Wright, J. Biol. Chem., 120, 641 (1937), 127, 137 (1939).

23. S. E. Darmon, G. B. B. M. Sutherland and G. R. Tristram, Biochem, J., 42, 508 (1948). See also R. C. Gore and E. M. Petersen, Ann. N. Y. Acad. Sci., 51, 924 (1949).

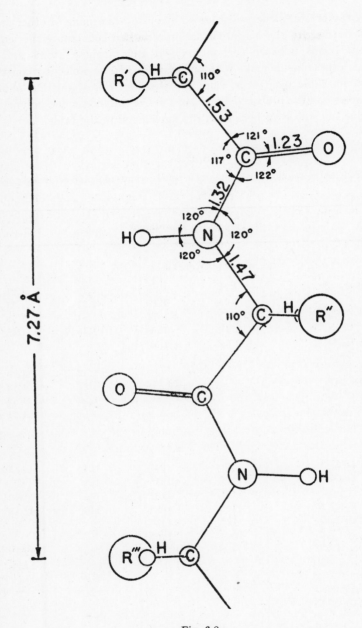

Fig. 6.8.

Probable dimensions of the fully extended polypeptide chain (Corey and Donohue)[24].

The X-ray crystal structure studies have also been made of other amino acids and the results give us important information concerning the lengths of covalent bonds between carbon, oxygen and nitrogen atoms and the angles which these bonds make with one another. These molecular constants are shown in Table 6.4 (taken from the paper of Corey and Donohue).[24] Fig. 6.8 represents the probable dimensions of the fully-extended polypeptide chain derived by these authors from the data listed in the table.

TABLE 6.4

Interatomic distances and bond angles found in some crystals of amino acids and peptides (Corey and Donohue[24])

| | L-Threonine | DL-Alanine | N-Acetyl-glycine | $\beta$-Glycyl-glycine |
|---|---|---|---|---|
| **A. Interatomic distances in Å** | | | | |
| Carboxyl C–O | 1.24 | 1.21 | 1.19 | 1.21 |
| | 1.25 | 1.27 | 1.31 | 1.27 |
| Carboxyl C–$C_\alpha$ | 1.52 | 1.54 | 1.51 | 1.53 |
| $C_\alpha$–N | 1.49 | 1.50 | 1.45 | 1.48 |
| $C_\alpha$–$C_\beta$ | 1.54 | 1.51 | | |
| $C_\beta$–C | 1.50 | | | |
| N-carbonyl C' | | | 1.32 | 1.29 |
| Carbonyl C'–O' | | | 1.24 | 1.23 |
| Carbonyl C'–$C_\alpha'$ | | | 1.50 | 1.53 |
| $C_\alpha'$–N' | | | | 1.51 |
| **B. Bond Angles** | | | | |
| Carboxyl O–C–O | 127° | 125° | 124° | 124.5° |
| Carboxyl O–C–$C_\alpha$ | 117° | 121° | 124° | 123° |
| | 116° | 113° | 112° | 112° |
| Carboxyl C–$C_\alpha$–N | 110° | 108° | 110° | 110.5° |
| Carboxyl C–$C_\alpha$–$C_\beta$ | 113° | 111° | | |
| N–$C_\alpha$–$C_\beta$ | 108° | 110° | | |
| C–N-Carbonyl C' | | | 120° | 122° |
| N-Carbonyl C'–O' | | | 121° | 125° |
| N-Carbonyl C'–$C_\alpha'$ | | | 118° | 114° |
| $C_\alpha'$-Carbonyl C'–O' | | | 121° | 121° |
| Carbonyl C'–$C_\alpha'$–N' | | | | 110° |

24. R. B. Corey and J. Donohue, J. Am. Chem. Soc., 72, 2899 (1950).

## 23. The Configuration of a Polypeptide Chain

Due to the internal rotation about single bonds as axes, a polypeptide chain can take various configurations, of which the simplest one would be the fully extended configuration:

EEEEEE....

Meyer and Mark[25] were the first to show that silk threads contain such extended chains: they were able to interpret the lattice of silk revealed by the X-ray diagrams as a covalent chain lattice in which the length of a peptide unit in the chain amounted to 3.5Å.

By stretching and rolling silk glands from the caterpillar Kratky and Kuriyama[26] were able to obtain ribbons of a higher degree of orientation. They found that the fiber period, $b$, is 6.95 ± .25Å and therefore, the half fiber-period, 3.5Å, is equal to the length of an amino acid residue. This period has been observed in other proteins.

If hair is allowed to soften in water or in dilute alkali, it can be stretched up to 100% and shows rubber-like elasticity: when the tension is removed, contraction takes place, and the hair returns approximately to its original length. Hair which has been steamed for a short period in the stretched condition, contracts to a length even shorter than the original length (supercontraction), while hair which has been heated for a longer period loses the power of contraction (permanent set).

These phenomena have been investigated in detail by Astbury, Street and Woods.[27] The most important result of this work is the discovery of the fiber-diagram of stretched hair, described by Astbury as the fiber-diagram of $\beta$-keratin. The characteristic differences between the diagram of the

25. K. H. Meyer, and H. Mark, Ber., **61**, 1932 (1928).

26. O. Kratky and S. Kuriyama, Z. physik. Chem., (B) **11**, 363 (1931).

27. W. T. Astbury and A. Street, Phil. Trans., (A) **230**, 75 (1931). W. T. Astbury and H. J. Woods, ibid., **232**, 333 (1933); W. T. Astbury, Chem. Weekblad, **33**, 777 (1936); J. Textile Inst., **27**, 281 (1936); Melliand Textilber., **76**, 1 (1935), etc. See also The Fundamentals of Fiber Structure. Oxford Press, London, 1933.

unstretched hair ($\alpha$-keratin) and that of the stretched hair ($\beta$-keratin) are as follows: The $\alpha$-diagram shows an arc on the meridian corresponding to a spacing of 5.1Å, while the $\beta$-diagram shows an arc corresponding to a spacing of 3.4Å: the $\alpha$-keratin diagram also shows a definite reflection on the equator corresponding to a spacing of 9.8Å. This reflection persists in the $\beta$-diagram, and in addition, a strong equatorial reflection appears, corresponding to a spacing of 4.5Å. Astbury concludes that $\beta$-keratin, like silk, contains stretched polypeptide chains, and that these arc held together laterally, in part by secondary valence in part by primary valence, e. g. the S–S-linkages. He interprets the spacings referred to above as follows:

$b = 3.4$Å $=$ the length of a peptide unit along the fiber-axis, equal to the length of a unit in silk-fibroin.

$a = 4.5$Å $=$ the spacing of the polypeptide chains from each other (backbone-spacing).

$c = 9.8$Å $=$ the spacing of the planes in which the polypeptide chains lie. (side-chain-spacing).

An interesting problem which naturally arises is the determination of the configuration of $\alpha$-keratin. The work of Astbury and his coworkers on $\alpha$-keratin has revealed that this configuration may be derived from fully extended $\beta$-keratin by assuming that the main chains contract to about one half of their original length, three peptide chains occupying 5.1Å. The side chain spacing is maintained at 9.8Å. The backbone spacing does not seem to be any longer present.

Some molecular models of $\alpha$-keratin have been proposed by Astbury and other investigators, including Huggins[28] and Shimanouchi and Mizushima.[29] One of the models proposed by the last-mentioned investigators is

BBBBB....

28. M. L. Huggins, Chem. Rev., **32**, 195 (1943).

29. T. Shimanouchi and S. Mizushima, Kagaku (Science) **17**, 24, 52 (1947); Bull. Chem. Soc. Japan, **21**, 1 (1948).

and the other is

EBBEBB....

Originally these two structures were proposed by Shimanouchi and Mizushima on the basis of the internal rotation potential including the *trans* planar configuration of the peptide $\overset{O}{\underset{}{>}}C-N\overset{}{\underset{H}{<}}$ bond: in other words all the movable atoms or groups in the main chain are in the stable positions of the internal rotation potential. It is very interesting that we could actually establish the existence of this folded form in the case of simple molecules such as acetylaminoacid N-methylamides as stated in Section 21.

The unit structure of the folded chain described above is similar to that proposed by Huggins, but there is a difference with respect to the position of the side chain. In consequence the Huggins model contains more *gauche* forms than the Shimanouchi-Mizushima model, and the former is considered to be less stable than the latter. The relation between these two models:

Huggins        and        Shimanouchi and Mizushima

can be considered to be similar to that between the polar and equatorial forms of methylcyclohexane or of monochlorocyclohexane, in which the equatorial form has been found to be much more stable than the polar form.

(See Section 14.)  In the present case the Huggins fold corresponds to the polar form and the Shimanouchi-Mizushima fold to the equatorial form.

The folded configurations proposed by Shimanouchi and Mizushima, as well as the extended configuration proposed by Meyer and Mark, satisfy the condition that all amino acid residues contained in the main chain are in L-forms.  There are many other configurations which satisfy this requirement and that of the internal rotation potential stated above.  In Fig. 6.9 are shown such possible configurations together with the extended and folded configurations referred to above.  (For these configurations special drawings have been made in Fig. 6.9 $a$ so that the spatial relations can be seen more clearly.)  $E$ and $B$ in the same figure are, respectively, the notations for the unit structures of the extended and the folded (or bent) configurations. These figures represent the projections of the molecular configurations on a plane perpendicular to one of the carbon bonds, where for the sake of simplicity all the bond angles of carbon and nitrogen are assumed to be tetrahedral, all the bond lengths of C–C and C–N to be equal and all the intramolecular rotational states about the C–C and C–N bonds as axes to be either the *trans* ($\theta = 0°$) or the *gauche* ($\theta = 120°$).

In an actual case the bond angles and the bond lengths will be different from these ideal values and, moreover, the intramolecular rotational state will easily be changed, to a certain extent, under the influence of the neighboring molecules.  Let us assume the bond lengths and bond angles of polypeptide chain to have the values equal to those found for glycylglycine:[30]

$$C_a\text{–}N = 1.48\text{Å} \qquad \angle\ C\text{–}C_a\text{–}N = 110°$$
$$N\text{–}C\ = 1.29\text{Å} \qquad \angle\ C_a\text{–}N\text{–}C = 122°$$
$$C\text{–}C_a = 1.53\text{Å} \qquad \angle\ N\text{–}C\text{–}O\ = 125°$$
$$C = O = 1.23\text{Å} \qquad \angle\ N\text{–}C\text{–}C_a = 114°$$

and the azimuthal angle of internal rotation to have the following values:

$$\theta\ (NH\text{–}CO\text{–}C_a\text{–}NH) = 140°$$
$$\theta\ (CO\text{–}C_\alpha\text{–}NH\text{–}CO) = 110°$$

Then the period along the chain of $BBB\ldots$ configuration is calculated as 5.14Å, which is in good agreement with the experimental value 5.15Å obtained by Astbury and Street.[27]  The N–H...O distance of this model is calculated as 2.52Å which is reasonable.  The presence of a two fold screw axis seems at first sight to be inconsistent with the fact that the 5.1 Å

30.  E. W. Hughes and W. J. Moore, J. Am. Chem. Soc., **71**, 2618 (1949).

EEEEE.....

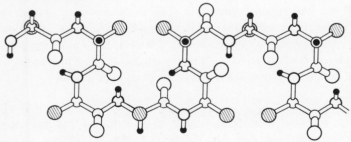

BBBBB.....

EBEBEB

BBEBBE.....

Fig. 6.9 a. Various configurations of polypeptide chain.

meridional reflection found in α-keratin is strong and well defined.[31] However, according to the experimental result obtained by Bamford, Hanby and Happey[32] a type of synthetic polypeptide gives an α-keratin diagram in which the 5.28Å polar arc is weak and diffuse. In view of this fact it is reasonable to propose the *BBB*... configuration as one of the possible structure of α-protein. It will easily be seen that *EBBEBB*... configuration can explain the reflection data more readily.

Let us explain various experimental results obtained for keratin and some other proteins from these polypeptide configurations.

(1) As already stated, the X-ray diffraction pattern of α-keratin can be explained by *BBB*... configuration (fiber period = 5.14Å) or by *BBEBBE*... configuration (fiber period = 10.3Å).

(2) In the α → β transformation the side chain spacing remains unchanged. This can be seen quite easily from the configurational change from *BBB*... to *EEE*... or from *EBBEBB*... to *EEE*... shown in Fig. 6.10. Both these changes do not affect the cystine bonds connecting the polypeptide chains laterally.

(3) The backbone spacing (4.5Å) of β-keratin exists no more in α-keratin This is quite understandable, because the *intermolecular* hydrogen bond which kept the neighboring polypeptide chains at the backbone spacing disappears to form *intramolecular* hydrogen bond.

(4) The supercontraction of wool can be explained if the keratin molecule which was originally in *BBB*... configuration is changed to take *EBBEBB*... configuration by the disconnection of the cystine chains. If the molecule was originally in *EBBEBB*... configuration, this may take one of the more folded forms shown in Fig. 6.9.

(5) The X-ray investigations show that the crystal structure of insulin,[33] excelsin,[34] etc. has trigonal symmetry. This is understandable if we consider that the molecule of such proteins is made of the ring configurations with trigonal symmetry such as shown in Fig. 6.9, or of the suitable superposition of these planar configurations. (The superposition may be caused through hydrogen bonds or through covalent bonds of side chains.)

---

31. W. T. Astbury, Nature, **164**, 439 (1949).
32. C. H. Bamford, W. E. Hanby and F. Happey, Nature, **164**, 138 (1949).
33. D. Crowfoot, Proc. Roy. Soc., London, (A) **164**, 580 (1938).
34. W. T. Astbury, S. Dickinson and K. Bailey, Biochem. J., **24**, 2351 (1935).

(6) It depends upon the nature of an amino acid residue, whether this residue takes the extended form or the folded form in the polypeptide chain. We have already seen that the different molecules of the type, $CH_3CONH$ $\cdot CHRCONHCH_3$, have different tendencies to assume the extended form

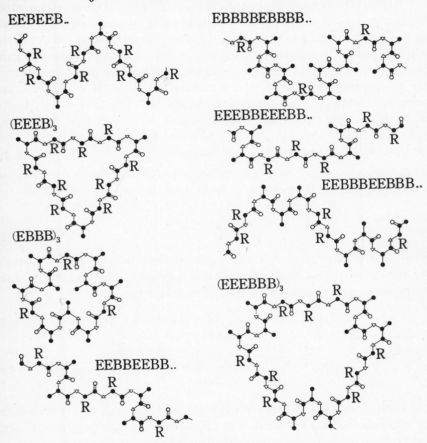

Fig. 6.9 b. Various configurations of polypeptide chain (continued).

or the folded form. (See Section 21.) There will be parallelism in configuration between the amino acid residues in polypeptides and acetyl aminoacid N-methylamides referred to above. Therefore, proline residue will show strong tendency to take the folded form.

The characteristic combination of the extended and folded forms of different aminoacid residues in a polypeptide chain will correspond to the

10

specificity of the protein. The importance of the order of aminoacid residues in a given protein will be related closely to the tendency of each residue in taking the extended form or the folded form.

(7) From the experimental results obtained for acetylaminoacid N-methyl-amide (Section 21), it will be seen that the two unit configurations, $E$ and $B$, do not differ much from each other in their internal energy and that the potential barrier to internal rotation over which $E$ and $B$ can pass into each other is not high. If, therefore, a denaturation involves a process of opening the polypeptide chain (i. e. the transformation from $B$ to $E$), we can understand why some proteins are denatured easily. We have already stated that the entropy change of acetylaminoacid N-methylamides on passing from the folded form into the extended form will explain reasonably the entropy change of some proteins on denaturation.

Let us next discuss the mechanical properties of wool.[29] For an elongation within two percent Hooke's law is found to hold. We have therefore,

$$E = \frac{1}{2} k \,(\Delta\, l)^2 \tag{6.3}$$

where $E$ is the energy change and $k$ the force constant. Let $y$ be Young's modulus referred to a single molecule and $l_0$ be an equilibrium length of an amino acid residue along the chain. Then $y$ is defined as

$$y = \frac{\partial E}{\partial \Delta\, l} \bigg/ \frac{\Delta\, l}{l_0} \tag{6.4}$$

From Eqs. (6.3) and (6.4) we have

$$y = k\, l_0 \tag{6.5}$$

The value of Young's modulus $Y$ in its ordinary sense can be obtained from the experimental relation between the tension $K$ and the elongation $\Delta\, L/L_0$:

$$Y = K \bigg/ \frac{\Delta\, L}{L_0} = \frac{10 \times 10^5}{0.02} = 5 \times 10^7 \,\text{g/cm.}^2$$

We can now put

$$\frac{\Delta\, l}{l_0} = \frac{\Delta\, L}{L_0}$$

and

$$y = Y A,$$

where $A$ denotes the molecular cross section. This can be calculated from the observed side chain spacing, 9.8Å, of $\alpha$-keratin and the backbone spacing which is assumed reasonably as 6Å. We then have

$$k = \frac{y}{l_0} = \frac{Y A}{l_0} = \frac{(5 \times 10^7 \times 980) \times (9.8 \times 6 \times 10^{-16})}{2.57 \times 10^{-8}} = 1.1 \times 10^4 \text{ dynes/cm}$$

(6.6)

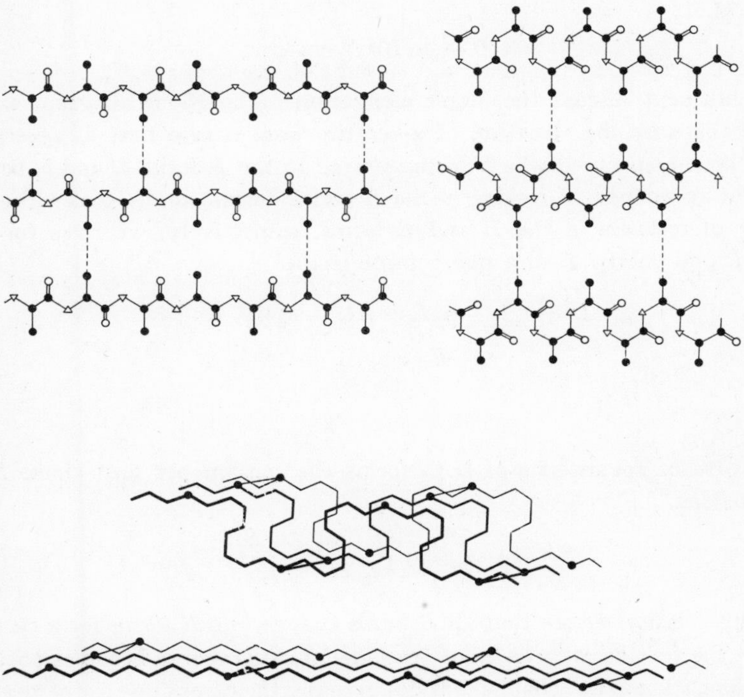

Fig. 6.10. The transformation from $\alpha$- to $\beta$-keratin. (The upper figure represents the transformation from $BBBBBB\ldots$ to $EEEEE\ldots$ and the lower figure the transformation from $EBBEBB\ldots$ to $EEEEEE\ldots$)

This value of $k$ is found to be quite reasonable, when compared with the force constant of a hydrogen bond.[35] (It is evident from the structure shown in Fig. 6.9 that the force which first resists the elongation of $\alpha$-keratin is the intramolecular hydrogen bond.)

The structure of $\alpha$-keratin may also be represented by $EBBEBB\ldots$ (see Fig. 6.9), for which the period along the main chain is calculated as

---

35. J. O. Halford, J. Chem. Phys., **14**, 395 (1946).

10.3Å.  As stated above, the reflection data for α-keratin can be explained by this model more simply than by that of $BBB$...  In this case the side chain (cystine bond, etc.) is directed upwards or downwards from the plane of Fig. 6.9 and therefore, the corresponding spacing (9.8Å) remains constant, when the chain is stretched out (see Fig. 6.10). If, therefore, we assume the backbone spacing to be 9Å, we can calculate the force constant $k$ just as in Eq. (6.6):

$$k = 0.84 \times 10^4 \, \text{dynes/cm.}$$

Let us next discuss the larger elongation.  The X-ray diagram shows that in this case the structure of α-keratin changes into that of β-keratin. Let $E_2$ be the energy of the $E$ form referred to the $B$ form, $l_1$ and $l_2$ be the length of an aminoacid residue in the $B$ and $E$ forms, and $n_1$ and $n_2$ be the number of residues in the $B$ and $E$ forms, respectively, we have for the length $L$ and energy $E$ of a polypeptide chain

$$L = n_1 l_1 + n_2 l_2 = n \, l_1 + n_2 \, (l_2 - l_1)$$
$$E = n_2 E_2$$

where

$$n = n_1 + n_2$$

In the case of coexistence of both forms the tension per unit chain $K$ is calculated as

$$K A = \frac{\partial E}{\partial L} = \frac{\partial E}{\partial n_2} \frac{\partial n_2}{\partial L} = \frac{E_2}{l_2 - l_1} \tag{6.7}$$

From this relation we see that all $B$ forms change into $E$ forms for a certain value of the tension, at which the tension-elongation curve becomes parallel to the elongation axis. That such is not actually the case is due to the neglect of entropy in the foregoing discussion.  The value of entropy will be very small in both extreme cases where all amino acid residues take $B$ or $E$ form, but in the intermediate case its value will be considerable, so that the curve inclines to the elongation axis to some extent.  If we put the value of tension $6 \times 10^5$ g/cm.$^2$ (corresponding to the middle point of the slope of tension-elongation curve observed by Speakman in the measurement for a relative humidity of 100 percent) into $K$ of Eq. (6.7) and put

$$l_2 - l_1 = 3.32 - 5.14/2 = 0.75 \text{Å}$$

we can calculate the energy difference between the $E$ and $B$ forms as:

$$E_2 = 370 \, \text{cal./mole (residue)}$$

Similarly for the structure of $EBBEBB\ldots$ we have

$$E_2 = 1800\,\text{cal./mole (residue)}$$

The energy difference between these two forms, is, therefore, small.

The elongation of wool cannot be explained by a single mechanism as the experiment of Bull shows.[36]   (The electron microscope experiment shows that keratin fiber has specific fine structures,[37] and therefore it may consist of complex micelles.)   Hence the foregoing discussion will not cover all the steps of elongation.   However, since X-ray diagrams show the structural change (i. e. $\alpha$ to $\beta$ change) after the elongation, the change of the intramolecular rotational state must be taken into account for the explanation of the mechanical property of keratin fiber.

It is interesting to note that the folded configuration, exactly the same as $BBB\ldots$ proposed by Shimanouchi and Mizushima, was also presented by Zahn[38] and by Ambrose and Hanby[39] quite independently.   The last-mentioned investigators made measurements with polarized infrared radiation of oriented films of poly-$\gamma$-methyl-L-glutamate cast from solution in $m$-cresol and concluded that the polypeptide chains take the $BBB\ldots$ configuration The evidence was based on the observation that the N–H stretching vibrations at 3300 and 3080 cm.$^{-1}$ show parallel dichroism, i. e. absorption is strongest when the electric vector of the incident radiation is parallel to the direction of orientation.   The N–H bonds in the $BBB\ldots$ configuration make angles of approximately 30° with the axis of the polypeptide chain (see Fig. 6.9).   Thus the change in dipole moment associated with the N–H stretching vibrations would have a larger component parallel to the chain than perpendicular to it, and the observed parallel dichroism would be understandable.   If the chain were in the $EEE\ldots$ configuration, absorption would be strongest for perpendicular radiation, just as we have seen in the case of crystalline acetylglycine N-methylamide (Section 21).   The results of the X-ray investigations by Bamford, Hanby and Happey[40] made in conjunction with the infrared measurements were compatible with the

---

36.  H. B. Bull and M. Gutmann, J. Am. Chem. Soc., **66**, 1253 (1944);  H. B. Bull, ibid., **67**, 533 (1945).

37.  E. H. Mercer, Nature, **159**, 535 (1947).

38.  H. Zahn, Z. Naturforschung, **2 B**, 104 (1947).

39.  E. J. Ambrose and W. E. Hanby, Nature, **163**, 483 (1949);  E. J. Ambrose, A. Elliott and R. B. Temple, Nature, **163**, 859 (1949).

40.  C. H. Bamford, W. E. Hanby and F. Happey, Proc. Roy. Soc. London, (A) **205**, 30 (1951).

*BBB*... configuration. Bamford and Hanby[41] also reported an interesting
result that some synthetic polypeptides in the *BBB*... configuration with
hydrocarbon side chains are soluble in nonpolar liquids, while after conversion
to the *EEE*... configuration, the polypeptides become completely insoluble
in nonpolar solvent. This is quite understandable, since in *BBB*... all
the hydrogen bonds are intramolecular, just as in the case of acetylproline
N-methylamide in carbon tetrachloride solution referred to in Section 21.

Ambrose and Elliott extended their infrared work to proteins and
obtained the following results. First they showed that in oriented films of
β-keratin (swan feather) the N–H bond is predominantly perpendicular to
the direction of extension of the polypeptide chain. This is in agreement
with the picture of the fully extended chain proposed for these proteins by
Astbury and Street.[42] However, for stretched films of myosin, tropomyosin
and for β-keratin, the dichroic properties of the NH bands at $3\mu$ are exactly
reversed showing that the maximum change in dipole moment for the NH
vibration takes place parallel to the chain. Therefore, the chain configuration
of these proteins should be different from that of β-keratin. They concluded
that these results can be explained reasonably by the *BBB*... configuration.[43]

Using the values of atomic distances and bond angles shown in Fig. 6.8,
and considering that the amino acid residues are equivalent, Pauling and
Corey[44] constructed two hydrogen-bonded helical configurations for the
polypeptide chain shown in Fig. 6.11. In these configurations the peptide
bonds have planar structure as in the case of *EEE*..., *BBB*..., etc. and
all the NH and CO groups are involved in hydrogen bonding in which the
nitrogen-oxygen distance is equal to 2.72Å and the vector from the nitrogen
atom to the hydrogen-bonded oxygen atom lies not more than 30° from the
N–H direction. (They considered by mistake that in the model proposed
by Shimanouchi and Mizushima and by Ambrose and Hanby, the planar
configuration of the peptide bond is not realized and the N–H...O angle
differs from a straight angle by 70°.*) For rotational angle 180° *the helical*
configurations may degenerate to a simple chain with all of the principal
atoms, C, N and O in the same plane.

---

\* As stated above the peptide bond in the Shimanouchi-Mizushima model has the
planar *trans* form and the N–H...O angle differs from a straight angle by 20 ∼ 30°.

---

41. C. H. Bamford and W. E. Hanby, Nature, **166**, 829 (1950).

42. W. T. Astbury and A. Street, Phil. Trans. Roy. Soc., **230**, 75 (1931).

43. E. J. Ambrose, A. Elliott and R. B. Temple, Nature, **163**, 859 (1949).

44. L. Pauling and R. B. Corey, Proc. Nat. Acad. Sci., **37**, 205, 235, 251, 256, **261**,
272 and 282 (1951).

Of these two helices, the 3.7-residue helix ($\alpha$-helix) is interesting in view of the experimental result recently obtained by Perutz[45] who found the 1.5Å reflection corresponding to the filber-axis length per residue.  It is probable that some proteins or synthetic polypeptide have this helical structure, but one goes too far, if one considers that only such helical

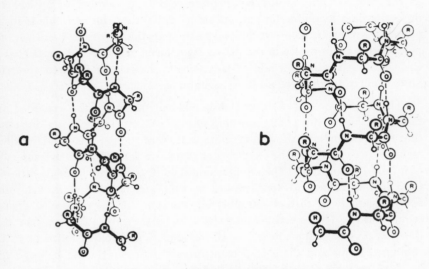

Fig. 6.11.  The helical configurations of Pauling and Corey.[44]
(a)  The 3.7-residue helix.   (b)  The 5.1-residue helix.

structures constitute an important part of the structure of proteins.  At any rate we cannot explain the specificity, the most interesting property of proteins, by such an uniform structure.  For the understanding of this important problem we have to consider the characteristic combination of different basic configurations such as the extended, folded and helical forms.*

---

* A systematic survey of polypeptide chain models which conform to the X-ray data has been made by L. Bragg, J. C. Kendrew and M. F. Perutz, Proc. Roy. Soc., (A) 203, 321 (1950).  A space-filling atomic model convenient for the investigation of polypeptide structure has been presented by Robinson et al.  See G. S. Hartley and C. Robinson, Trans. Farad. Soc., 48, 847 (1952); C. Robinson and E. J. Ambrose, ibid., 48, 854 (1952).  Recently Shimanouchi and Mizushima derived a general formula expressing the configuration of a polypeptide chain in terms of atomic distances, bond angles and internal rotation angles.  See T. Shimanouchi and S. Mizushima, J. Chem. Soc. Japan, 74, 755 (1953); S. Mizushima, Advances in Protein Chemistry, IX.

---

45. M. F. Perutz, Nature, 167, 1053 (1951).

## Summary of Chapter VI

One of the methods of attacking the structural problem of proteins is to study the structure of aminoacids, amides and other simple substances closely related to proteins. In this chapter the results of such studies are described beginning with that for N-methylacetamide which can be regarded as the simplest unit of the polypeptide chain. A number of optical and dielectric data available for this substance show that the molecule is in the planar *trans* form. This result suggests that the peptide bond in the polypeptide chain is also in the planar *trans* form. This view is supported by the fact that the main chain of polypeptide shows near infrared spectrum quite similar to that of N-methylacetamide.

The molecules with two peptide bonds can take both the extended and the folded forms with stable positions of internal rotation potential including that corresponding to the planar configuration of the peptide bond. Experimentally we can prove the existence of these two forms by the measurement of infrared absorption. The equilibrium ratio of these two forms is different from one substance to another and in the extreme case of acetylproline N-methylamide in carbon tetrachloride solution, all the molecules exist in folded forms. The experimental results for these substances suggest that the polypeptide chain can be in the extended and folded forms or in the combined forms and the α to β transformation of keratin can be explained as the structural change from the folded to the extended form. The X-ray data of silk, stretched wool and some other fibers can be accounted for by the extended form, while those of natural wool and some synthetic fibers by the folded form. The results of the measurements with polarized infrared radiation and of solubility experiment made for some protein and synthetic fibers are explained from the same point of view. A helical form has also been presented as a configuration of some α-proteins and synthetic polypeptides. The specificity of a protein can be explained by considering specific combinations of different configurations referred to above.

# PART II

In Part I we have presented a rather general description of the problem of internal rotation. The details of the theoretical aspects of the problem, especially those of normal vibrations as well as the experimental methods remaining to be described will be discussed in this part.

As in Part I many sections of Part II contain results of research work especially developed in our laboratory, but the author believes it necessary to include certain sections well known to the specialist, because this part will also be of interest to the average chemist whose main endeavor lies in other directions, but who wishes to have a broad acquaintance with the general field of structural chemistry.

# CHAPTER I

# Principles and Experimental Methods of Structure Determination

## 1. Infrared Absorption

The internal energy of a molecule is restricted to a series of discrete levels or stationary states and a transition of the molecule from one of these stationary states to another gives rise to the emission or absorption of radiation. For the interpretation of the spectrum due to any molecule we must, therefore, try to determine its energy states and the "selection rules" governing the transitions between them. Now the internal energy of a molecule comes partly from the motions of electrons, partly from the mutual vibrations of atomic nuclei, and partly from the rotation of the molecule as a whole. The interaction among these three motions is in most cases negligible and these three types of energy will be quantised separately, giving rise to three kinds of molecular spectra — electronic, vibrational, and rotational. Since the electronic energy states are normally of the order of a few electron-volts (or hundred kcal./mole) above the ground state, electronic spectra are confined almost entirely to the visible and ultra-violet regions of the spectrum. However, as the vibrational energy levels lie only a few tenths of a volt, and the rotational states about a hundredth of a volt, above the ground state, the spectra due to changes in these fall in the infrared, the former being confined roughly to the region between $1\mu$ and $30\mu$, the latter ranging from $15\mu$ to $500\mu$.

Most of the spectroscopic description given in part I is concerned with the vibrational spectra and, therefore, the experimental methods of this section will mainly deal with measurements in the wave-length range from $1\mu$ to $30\mu$.

The infrared spectrometer used in work in this field consists essentially of: (1) a source emitting a continuous range of wave lengths desired: (2) a dispersing means to spread out this radiation in order to provide narrow wave-length bands at accurately known wave-length positions: (3) a means for interposing a sample of suitable thickness into the path of this radiation: (4) a detector and amplifier to measure accurately the intensity of radiation in each narrow band.

Today several commercial spectrometers such as the Baird, the Perkin-Elmer and the Beckman instruments are available. They are very convenient and efficient, answering the increasing demand for use in inorganic and organic laboratories as well as those in chemical industry. However, for the laboratories specializing in this field it is often necessary to construct instruments suitable for their special purposes. The apparatus which will be described in the following belongs to a common type, but is worth while

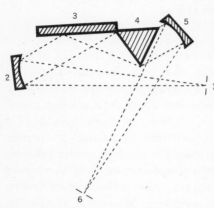

describing, because almost all the essential parts have been made by ourselves and, therefore, the apparatus may be of interest to those who wish to make infrared instruments or a part of them according to their individual needs.

Fig. 1.1. Reflecting monochromator.

*A. Monochromator.* Figure 1.1 shows the schematic drawing of the reflecting monochromator used in our laboratory. The beam of light entering through the slit (1) falls first on the concave mirror (2). The parallel beam which emerges from this is reflected by the plane mirror (3) on to the prism (4) which splits it up into a spectrum. Finally the light falls on the second concave mirror (5) which projects a spectrum in the plane of the exit slit (6). The plane mirror (3) and the prism (4) are mounted together on a rotating table. The prism, which can be easily changed for another (glass, quartz, rook salt and sylvine prisms), is at every setting in the position of minimum deviation for the wave-length which is transmitted by the exit slit (6).

*B. Prisms.* As to prisms for use in the monochromator, there has been great improvement in the availability of synthetic crystals. Nowadays it is not difficult to prepare such crystals in the ordinary chemical laboratory and even to make a prism with them. In the following the essential feature of the process adopted in our laboratory will be described.[1]

The method is that developed by Kyropoulos,[2] in which the crystal is drawn from the melt. A seed crystal is held in a chuck forming the lower

---

1. K. Kuratani, Rep. Rad. Chem. Res. Inst. Tokyo Univ., **5**, 25 (1950).
2. S. Kyropoulos, Z. anorg. Chem., **154**, 308 (1926).

end of a vertical metal tube which can be cooled by water and is supported over the mouth of a vertical furnace, inside which is the crucible containing the molten salt.  The seed is then lowered without water in the metal tube so that the lower end enters the liquid and it is allowed partially to melt so as to form a fresh surface.  After 1 ∼ 2 minutes the metal tube is cooled by running water so that crystallization on the seed commences.  Then the crystal is lifted up 1 ∼ 2 mm every fifteen minutes.  Repeating the process 10 ∼ 20 times, we can obtain a single crystal of a height of 5 ∼ 10 cm.  Then the crystal is lifted clear of the residual melt and is cooled down to ordinary temperature in 120 ∼ 40 hours.

The synthetic crystal thus prepared can be cut and polished on a pitch surface which is channeled by cutting it with a hot knife.  Miyazawa has made several prisms which have been used satisfactorily in our infrared instrument.  Besides the synthetic crystals referred to above a solid solution of thallous bromide and iodide, which is commercially available, is excellent for use for prisms in infrared spectrometers.  While the material is actually opaque in the blue region of the visible spectrum (to wave-lengths shorter than 5,000Å), it is transparent out to $40\mu$ in the infrared: it, therefore, makes available the range of wave-lengths from 25 to $40\mu$.  (Up to $25\mu$ one can work with potassium bromide prisms.*)

*C. Thermoelement and galvanometer.*  For longer wave-lengths the radiant energy is measured in our laboratory by a vacuum thermoelement made of Bi and Bi plus 5% Sn.  This is prepared by the Taylor process as follows.  The metals are melted and sucked up into a thin-wall capillary tube of soft glass.  This tube, containing the metal as a core, is heated in a small electric furnace and drawn out.  Then the glass is removed from the composite fibers with hydrofluoric acid diluted with a little water to suppress fuming.  A tiny piece of Wood's metal fused to the junction by radiation and wetted with flux (a solution of pure $ZnCl_2$ in distilled water) facilitates attaching the receiver which is made of thin Cu foil blackened by Japanese ink, "Sumi".  The element is enclosed in a small glass vessel with the KCl window.  The thermoelectric current generated by the emergent radiation is measured by a loop galvanometer, the sensitivity of which is $8 \times 10^{-8}$ volt for one scale division of the eyepiece.  The internal resistance of the galvanometer is almost equal to that of our thermoelement (several ohms) and the period is about 0.5 sec.

---

* Recently cesium bromide prism has been used in place of the thallous bromide and iodide prism.

In measurements with small amounts of sample or by polarized infrared radiation, where the available radiation is limited, it is often necessary to increase the sensitivity of the measurement. For this purpose an apparatus was made by Shimanouchi and Miyazawa in which the deflection of the loop galvanometer is amplified by a photoelectric method (Fig. 1.2). The light

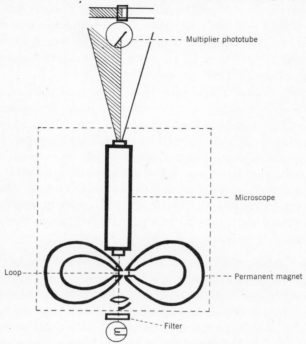

Fig. 1.2. Photoelectric amplifier.

from a 200 W projector lamp with its infrared part removed by a filter, is focussed on the loop, whose image is projected through a microscope ($\times$ 640) onto the photocathode of a multiplier phototube. By darkening one side of the image of the loop, we can make a small shift of the image greatly affect the amount of the output current of the multiplier phototube. To avoid the error arising from nonlinearity between the displacement of the loop and the current change, a double beam method was used. The energy difference between the beam passing through the sample and the reference beam is measured and a neutral wedge is inserted into the reference beam so as to make the difference zero, and the absorptivity of the sample is obtained from the attenuation by the neutral wedge.

An outstanding detector development of recent years is the development by Cashman of photocells based on lead sulfide.* Other substances, e. g. lead telluride have also been investigated. The lead sulfide cells extend the range of photocell technique to $3.5\mu$, and the lead telluride to about $6\mu$. They are not only much faster in response than thermocouples, but also more sensitive.

An even more rapid, and apparently more sensitive, detector than the thermocouple has also been developed by Golay. This pneumatic detector, like a thermocouple, is nonselective and is thus applicable to the detection of radiation of all wave-lengths.**

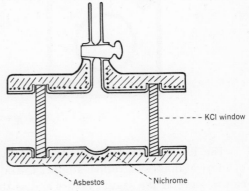

Fig. 1.3.   Gas absorption cell.

*D. Absorption cells and sampling.* The gas absorption cell is made of a glass cylinder to which KCl windows are sealed by glyptal as shown in Fig. 1.3. If the sample is in the liquid state at room temperature, it is evaporated in the cell by means of heated nichrome wire wound around the glass cylinder. In order to avoid the condensation of the vapor on the windows as well as to keep the temperature uniform throughout the cell, short glass cylinders heated electrically are attached in front of the windows. This cell has been found quite satisfactory for obtaining the vapor spectrum at any desired temperature.

Liquid sampling can be done as follows: If the liquid is nonvolatile and only a survey spectrum is desired, it may be spread on a KCl plate and a cover plate placed on top. The assembly is held in a cell holder. Thickness variation can be attained by separating the plates with $U$-shaped metal foil spacers of known thickness. The cell for volatile liquids (see Fig. 1.4) consists of two KCl plates separated by a Hg-amalgamated lead spacer and the assembled sandwich clamped firmly in a cell holder to give a tight cell.

* As to the method of manufacture see L. Sosnowski, T. Starkiewicz and O. Simpson, Nature, **159**, 818 (1947); M. Eguchi, A. Murakami and Y. Oshima, *Seisankenkyu*, **3**, 89 (1949); M. Tsuboi, J. Chem. Soc. Japan, **72**, 616 (1951).
** The Golay detector is commercially available. (The Eppley Laboratory, INC.)

Filling and emptying the cell can be accomplished by means of the holes drilled in one plate.

Solid samples are studied as melted films, if possible. When suitable melted films are not obtained, the sample is dispersed in nujol and the paste is spread on the plate to the desired thickness. This technique precludes sample data in the nujol C–H absorption region at 2900 cm.$^{-1}$, 1450 cm.$^{-1}$ and 1360 cm.$^{-1}$ but the rest of the spectral range is quite available.*

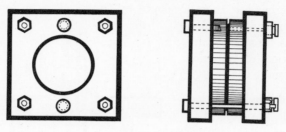

Fig. 1.4. Liquid absorption cell.

The low temperature cell designed by Miyazawa consists of an outer cylindrical brass can, rubber stopper to seal this can, and an inner cylindrical can to serve as the sample holder and as the reservoir for cryogen (Fig. 1.5). The lower part of the outer can has KCl windows sealed with compound in order to pass the infrared radiation through the cell. To keep the temperature of the KCl windows slightly above the room temperature (otherwise the KCl windows would become fogged), a heated nichrome wire is wound around the outer can just above the windows. From the stopper a brass cylinder is suspended by means of a narrow brass tube which serves as cryogen inlet. The inner can has a hole for passing the light beam. The sample sandwiched between two KCl plates, is held in this position. To measure the temperature of the sample a copper-constantan thermocouple is held in contact with the KCl plate. After the whole is assembled, the cryogen (liquid air or dry ice-acetone mixture) is introduced into the inner can, and the space between the outer and inner cans is evacuated so as to reduce the convection loss and to prevent the deposition of ice on the specimen.

---

* An interesting technique for the measurement of solid samples has been introduced by Sister Miriam Michael Stimson at Adrian, Michigan. A powder of potassium bromide is mixed with a small quantity of the sample and is pressed under high pressure to form a transparent disk. When no decomposition takes place, the disk shows characteristic absorptions of the sample.

*E. Use of polarized radiation.* Another recent improvement in technique is the use of polarized infrared radiation. The idea of polarized infrared is not new, and one means of polarizing such radiation has been known for a long time, polarization by reflection from selenium, as was first proposed by Pfund in 1906.[3] Plane selenium mirrors have been used in our laboratory with an angle of incidence of 71°, as suggested by Czerny,[4] to obtain polarized radiation of wave-length greater than 2 micron. Another method for obtaining the polarized radiation is to use flat sheets of silver chloride. An assembly consisting of several selenium films is also very suitable as a polarizer (Elliot, Ambrose and Temple).[5]

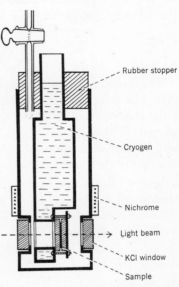

Rubber stopper

Cryogen

Nichrome

Light beam

KCl window

Sample

Fig. 1.5. Low temperature cell for the solid.

The sample is examined successively with the direction of orientation of the crystals parallel or perpendicular to the direction of vibration of the electric vector in the polarized incident beam. Differences in relative intensity of the absorption bands are found with many substances as the sample is rotated and it is possible from this measurement to determine the geometrical structure within the molecule, as has been described in Part I.

Except for special cases the amount of light transmitted by a sample is governed by Beer's law*

$$\ln\left(\frac{I}{I_0}\right) = -\varkappa N l.$$

$I_0$ is the radiation incident on, and $I$ the radiation transmitted by, the sample at the frequency $\nu$, $\varkappa$ is the molecular absorption coefficient of the sample material at the frequency $\nu$, $N$ is the number of molecules of a given material per c. c. and $l$ is the thickness of the absorbing layer of the sample.

---

* See Eq. (2.10), Part I.

3. A. H. Pfund, Johns Hopkins Univ. Circ., **4**, 13 (1906).
4. M. Czerny, Z. Physik, **16**, 321 (1923).
5. A. Elliot, E. J. Ambrose and R. B. Temple, J. Opt. Soc. Am., **38**, 212 (1948).

The value of $\varkappa$ will be appreciable only when $\nu$ is an infrared-active frequency. In general, $\varkappa$ will have a larger value for a more polar compound, as we have seen in the case of the N–H absorption band. (As explained in Section 21, Part I, the molecular absorption coefficient of the N–H absorption band becomes stronger, as the bond acquires more polar nature due to resonance.) The value of $\varkappa$ is fairly insensitive to pressure and, except in special cases, is practically independent of temperature.

## 2. The Raman Effect

The Raman effect is a phenomenon depending on the scattering of light which has been studied by many investigators. The scattered light known as the Rayleigh radiation has the same frequency as the incident radiation but careful examination shows that in addition to the Rayleigh radiation there is a small amount of scattered light which has frequencies different from that of the exciting light, but which cannot be explained as fluorescence.

In 1928 Raman in India (and almost at the same time Landsberg and Mandelstam in Russia) found that when certain molecules are irradiated with monochromatic light of frequency $\nu$, the scattered light contains several new frequencies, $\nu - \nu_1$, $\nu - \nu_2$... (called Stokes lines), and others $\nu + \nu_1$, $\nu + \nu_2$,... (called anti-Stokes lines), symmetrically disposed about the exciting frequency $\nu$, where $\nu_1, \nu_2$... are of the order of infrared frequencies and, in fact, are often found to coincide with known infrared absorption frequencies of the scattering molecule. Qualitatively, such an effect had been predicted in 1923 by Smekal on the simple theory that an incident light quantum might give up a part of its energy to excite a molecule, with the result that the scattered light quantum would be diminished in frequency by an amount corresponding to the energy given to the molecule. If the molecule were in an excited state, it might give up energy to the incident light quantum, thus increasing the frequency of the scattered quantum. The former process would account for the Stokes lines and the latter for the anti-Stokes lines, giving a very satisfactory explanation of why the latter are much weaker than the former since the number of molecules in excited levels must decrease exponentially according to a Boltzmann factor: i. e. for the Raman line of the frequency $\nu + \nu_1$, the number of molecules in the excited state is reduced by factor $e^{-\frac{h\nu_1}{kT}}$. A more accurate theory of the effect had also been given as early as 1925 in a paper by Kramers and Heisenberg of the dispersion of light. The essential feature of their argument is that the incident light quantum interacts with the molecule, raising it to some unstable upper level, from which it immediately drops back, usually to the ground state, but sometimes to an excited vibrational or rotational level. In the former case, the frequency of the incident light is unaltered (corresponding to the Rayleigh scattering), while in the latter case new frequencies will be observed (corresponding to Raman scattering).

At first it was thought that the Raman lines have the same frequencies as the infrared absorption peaks, but this was soon shown to be wrong. The fact that infrared absorption involves only one transition between two states of the molecule, whereas in the production of a Raman spectrum there is a double transition would make one expect a difference in the selection rules operative in the two cases.

There is no need to explain in detail the quantum theory of the Raman effect, but a short description based on classical mechanics is necessary for the explanation of the various spectroscopic problems with which we have dealt in Part I. For the sake of simplicity let us consider that the scattering particle is a diatomic molecule with one fundamental frequency $v_1$. If it is irradiated with monochromatic light of frequency $v$, the electrons are periodically displaced in the rhythm of the incident light and a dipole moment is induced in the molecule. Let the electric vector of the incident light be:

$$E = E_0 \cos 2\pi v t, \tag{1.1}$$

then the induced moment,

$$m_i = \alpha_0 E_0 \cos 2\pi v t, \tag{1.2}$$

where $\alpha_0$ is the polarizability and is a measure of the ease with which the negative charge distribution of the molecule may be deformed. The secondary waves which are emitted by the molecule are of the same frequency as that of the incident light and these constitute the Rayleigh radiation.

Now let the molecule vibrate along the line joining the nuclei. The polarizability will vary with the internuclear distance and for a small displacement $x$

$$\alpha = \alpha_0 + \alpha_1 x \tag{1.3}$$

and if the vibration is simple harmonic

$$x = x_0 \cos 2\pi v_1 t. \tag{1.4}$$

The induced moment for the vibrating molecule now becomes

$$m_i = \alpha E = (\alpha_0 + \alpha_1 x) E_0 \cos 2\pi v t$$

$$= \alpha_0 E_0 \cos 2\pi v t + \frac{1}{2} \alpha_1 x_0 E_0 \{\cos 2\pi (v + v_1) t + \cos 2\pi (v - v_1) t\} \tag{1.5}$$

The scattered light now consists of the Rayleigh radiation and two new frequencies, $\nu \pm \nu_1$. Through Eq. (1.5) we see that a vibration cannot appear as a Raman line unless

$$\alpha_1 \neq 0$$

In other words, the change of polarizability during a molecular vibration is responsible for the Raman effect. This selection rule is quite different from that of infrared absorption in which a vibration giving rise to a changing electric moment is active. This is the picture of classical mechanics corresponding to the quantum theory stated above. Thus the vibration of a homonuclear diatomic molecule, such as $H_2$, being symmetrical about its center, can be observed in the Raman effect but not in absorption, since such a vibration can cause change in polarizability but not in electric moment. More generally, if any polyatomic molecule has a center of symmetry, then those vibrations which are symmetrical with respect to the center will be observed only in the Raman effect and not in the infrared absorption. This

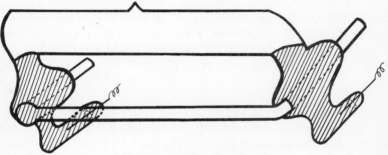

Fig. 1.6. Quartz mercury lamp.

is an important theoretical conclusion to which we have often referred in the spectroscopic discussions of Part I. However, we do not intend to make any rigorous derivation of the selection rule which would be beyond the scope of this book.

The intensity of a Raman line is only a few hundredths or thousandths of the intensity of Rayleigh scattering, which itself is an even smaller fraction of the light intensity incident on the sample. Therefore, very high intensity light must be effectively concentrated on the sample. The most common source of light is the mercury arc. Today various kinds of commercial mercury lamps are available, but it is necessary to choose such kinds with a high ratio of Hg-4047 or 4538 intensity to continuous background intensity, because Raman lines tend to be masked by the background. In our laboratory we use quartz mercury lamps, which operate at low vapor pressures and the electrodes of which are cooled with circulating water (Fig. 1.6). Such lamps give sharp and intense mercury lines with very low continuous background. The lamp is placed at one focus of an elliptic reflector (coated with MgO, $MgF_2$, etc.) and at the other the Raman tube containing the liquid sample, surrounded by a double-walled glass mantle through which water of any desired temperature can be circulated. (Fig. 1.7.)

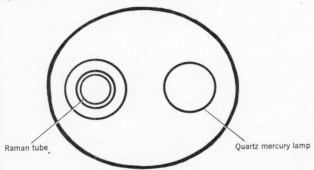

Raman tube

Quartz mercury lamp

Fig. 1.7. Quartz mercury lamp and Raman tube in position.

For photographic Raman work, the speed of a spectrograph depends on the relative aperture of its camera system. Even with high intensity source equipment, an aperture approaching $f/4$ is desirable for most Raman work. With a combination of such source equipment and a fast spectrograph, spectra of typical compounds can be registered in a few minutes. This allows for taking weaker Raman spectra or spectra of mixtures without resorting to excessive exposure times.

Most Raman work requires the use of high-speed film that has rather poor resolving power. Therefore, good linear dispersion of the spectrograph is required. This is also desirable so that position measurements can be made accurately.

The combined requirements of high speed and good linear dispersion are the most important requirements for a Raman spectrograph. To meet these requirements spectrographs with optical systems of large cross-sectional area (diameter of lenses 12 cm., height of prisms $12 \sim 15$ cm.) have been constructed in our laboratory. Each one of these spectrographs contains at least two 60° prisms of highly dispersive, extra dense, flint glass.*

---

* These prisms and lenses were manufactured by the Japan Optical Company.

Direct-recording arrangements for Raman spectra have been constructed by a number of workers, including Rank[6] and Miller, Long, Woodward and Thompson.[7] Commerical scanning equipment is made by Applied Research Laboratories in the United States, and another by Hilger and Watts, Ltd. in England. In the latter, the incident radiation is chopped at 22 c/s by a sector disc maintained by an oscillator circuit. The multiplier signal is a. c. amplified and applied after rectification to the pen of a ratio recorder.

Filters are usually interposed between the source and the sample in order to remove high energy light which might cause photodecomposition, isolate a single exciting line, and remove the continuous spectrum in the region occupied by the Raman lines. Especially in the case of the crystal powder method, it is necessary to make the light source monochromatic by the use of filters.

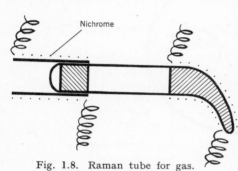

Fig. 1.8.   Raman tube for gas.

For Hg-4358 excitation a saturated solution of sodium nitrite is used to remove ultraviolet and violet light; concentrated carbon tetrachloride solution of iodine is used for excitation Hg-4047; a dilute solution of the same substance and a solution of praseodymium nitrate is used to remove the continuum on the longer wave-length side of Hg-4358.

Many substances discussed in Part I are in the liquid state at ordinary temperatures and we have been interested to take their Raman spectra in the vapor state. A Raman tube used by us for this purpose is shown in Fig. 1.8. It has a capacity of about 1500 c. c. A calculated amount of the sample is introduced in the tube, so that it forms a vapor of $4 \sim 5$ atmospheres when heated at the temperature at which the measurement is made. The heating is effected by the nichrome wires wound around the head and the tail of the Raman tube. In the case of such a substance as 1,2-dichloroethane the heat radiated from the mercury lamp is sufficient to keep it in the vapor state, nevertheless it is desirable to use the electric heating in order to avoid the condensation of the vapor on the wall at the head or the

6. D. H. Rank, R. J. Pfister and P. D. Coleman, J. Opt. Soc. Am., **32**, 390 (1942).

7. C. H. Miller, D. A. Long, L. A. Woodward, and H. W. Thompson, Proc. Phys. Soc., **62 A**, 401 (1949).

tail of the tube. A lens is placed between the tube and the spectrograph in such a position that the beams parallel to the axis of the tube are focussed upon the slit.

Raman tubes used by us for the measurement of liquids at different temperatures are shown in Figs. 1.9 and 1.10. In the case of a very small quantity of a liquid sample to be measured at room temperature we have constructed a special vessel described in the following in addition to the ordinary cylindrical tube in miniature.[8]

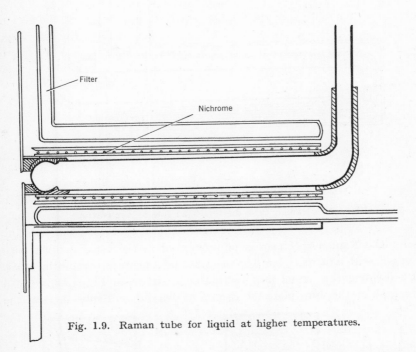

Fig. 1.9. Raman tube for liquid at higher temperatures.

If we place an electric lamp on the focal plane of the camera lens of a spectrograph (i. e., on the plane of the photographic plate), then the light passing through the spectrograph is emitted from the slit $S$ as monochromatic light. A lens $L$ is placed in front of the slit at such a distance from $S$ that its image formed at $I$ of Fig. 1.11 is about twice as large as its original size. In the same figure the broken lines indicate the thin beam of light emitted from the slit. A glass plate cut off in part in the form of this beam and held

8. S. Mizushima, T. Shimanouchi and T. Sugita, J. Am. Chem. Soc., **72**, 3811 (1950).

between two other plates is used as the Raman vessel which is placed in the position shown in Fig. 1.11 in which the shaded part is blackened (or covered with black paper) in order to avoid the entry of unnecessary light into the

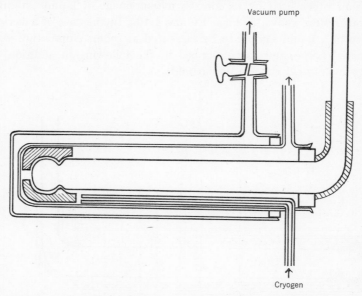

Fig. 1.10.  Raman tube for liquid at lower temperatures.

vessel.  The Raman spectrum is photographed in the usual manner with this vessel, with a mercury lamp on one side and a plane mirror on the other.

By dividing this vessel into two parts as shown in Fig. 1.12, we can photograph on the same plate the spectra of two different substances at the

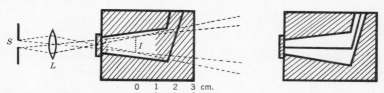

Fig. 1.11.  Raman vessel for a small quantity
of liquid.

Fig. 1.12.  Raman vessel
divided into two parts.

same time and thus the comparison of the two spectra can be made easily and accurately.  We can also make polarization measurements with this vessel by covering one part with a polaroid transmitting the light vibrating parallel to the spectrographic axis and the other with a polaroid which transmits light vibrating perpendicular to the axis.

As stated in Section 4, Part I the experimental technique of observing the spectral difference between the liquid and solid states provides the simplest method of detecting rotational isomerism in the liquid state. For this purpose a convenient apparatus for observing the low-temperature

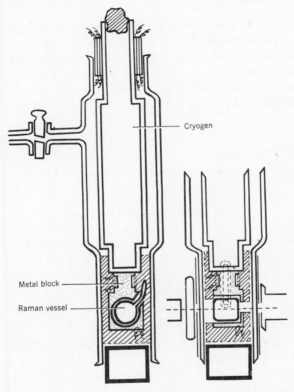

Fig. 1.13. Raman vessel for solid at low temperatures.

solid sample, which is in the liquid state at ordinary temperature, has been constructed by Ichishima in our laboratory. As shown in Fig. 1.13, the sample in the glass cell is cooled by liquid air or dry ice-acetone mixture through the conduction by the metal block.

The measurement of depolarization factors can be made in two different ways: one involving a single exposure and the other involving two exposures. In the former method, the substance under investigation is illuminated in a single plane by parallel radiation, and the scattered light is observed in a direction strictly perpendicular to the incident light. The beam is split into two components by the Wollaston prism or a split-field polaroid, and recorded simultaneously one above the other on the same photographic plate. The depolarization factor $\rho$ is defined as:

$$\rho = I_\perp / I_{||}$$

where $I_\perp$ and $I_{||}$ are, respectively, the intensities of the components of scattered light with electric vectors vibrating in the directions perpendicular and parallel to that of the incident light. Preferential reflection of the polarized beams at optical surfaces must be avoided. This can be done by using half-wave mica plates, or by placing the resolving polaroids in such a manner that their line of contact is 45° from the vertical axis of the spectrograph.

In the double-exposure method, the two polarized beams are obtained in two successive exposures, with a rotation of some part of the appartus (lamp, nicol prism, or polaroid) through 90° between exposures.

### 3. Dielectric Constant

Molecules are composed of positive nuclei and negative electrons. If the center of action of the positive parts coincides with that of the negative parts, the molecule is non-polar and has no permanent dipole moment: examples are $H_2$, $O_2$, $CH_4$, $CCl_4$, $C_6H_6$, etc. If the two centers do not coincide, the molecule is polar and in the electric field will tend to arrange itself with the negative end of the permanent dipole toward the positive pole. The magnitude of the dipole moment of a molecule is the order of $10^{-18}$ electrostatic units or $D$ (Debye unit).

The methods employed for measuring dipole moments depend almost wholly on the determination of the dielectric constant, $\varepsilon$, which is the ratio of the capacity of the condenser filled with the medium to its capacity when evacuated, or the square of the refractive index, provided waves of the same frequency are used to measure both the dielectric constant and the refractive index. The quantity called molecular polarization $P$ of a substance of molecular weight $M$ and density $d$, is related to the dielectric constant by the equation:

$$P = \frac{\varepsilon - 1}{\varepsilon + 2} \cdot \frac{M}{d} \qquad (1.6)$$

This corresponds to the formula for the molecular refraction of a medium of refractive index $n$:

$$R = \frac{n^2 - 1}{n^2 + 2} \cdot \frac{M}{d} \tag{1.7}$$

If we have a mixture of two liquids of molecular polarizations $P_1$ and $P_2$, molecular weights $M_1$ and $M_2$, present in the molecular fractions $x_1$ and $x_2$ we can express the mean molecular polarization of the mixture $P_{1,2}$ by writing the foregoing relation in the form:

$$P_{1,2} = \frac{\varepsilon - 1}{\varepsilon + 2} \left( \frac{x_1 M_1}{d_1} + \frac{x_2 M_2}{d_2} \right) = x_1 P_1 + x_2 P_2 \tag{1.8}$$

In this way we can determine the molecular polarization of a substance either alone or as the constituent of a mixture.

In the case of a nonpolar substance the value of $P$ is independent both of the concentration and of the temperature. In the case of a polar substance, however, it is found that the value is affected by both of these variables. The change with concentration is due to the influence of the dipoles on one another, which naturally increases as they get nearer. Accordingly in order to use Equations (1.6) and (1.8), we have to eliminate the mutual influence by removing the dipoles sufficiently far from one another, either by measuring the substance as a vapor or by taking its dilute solution in a non-polar solvent.

In order to understand the effect of temperature, we must consider the nature of the polarization in more detail. The polarization of the molecules consists of the following three parts:

(1) The electrons will be displaced with respect to the nuclei towards the positive pole of the electric field. This is the part due to electronic polarization, $P_E$.

(2) As a result of this deformation of the electronic clouds, the nuclei themselves will be displaced with respect to one another: this gives the atomic polarization, $P_A$ which is usually small compared with $P_E$.

(3) These two effects give the molecule a temporary dipole moment, but if the molecule has a permanent dipole moment, the external field will tend to orient it along the direction of the lines of force, giving rise to the orientation polarization, $P_0$.

Since the orientation will be opposed by the thermal motion of the molecule, it should diminish as the temperature rises. The exact relation was deduced by Debye,[9] according to whom the total molecular polarization is given by:

$$P = \frac{\varepsilon - 1}{\varepsilon + 2} \frac{M}{d} = \frac{4\pi}{3} \cdot N \left( P_E + P_A + \frac{\mu^2}{3 k T} \right) \qquad (1.9)$$

where $N$ is the Avogadro number, $T$ the absolute temperature, and $k$ the Boltzmann constant and $\mu$ the dipole moment.

Therefore, if we measure $P$ for a polar substance at a series of temperatures, we can calculate the value of $\mu$, since $P_E + P_A$ is not dependent upon temperature. This method, in which the electronic and the atomic polarizations are eliminated by making measurements at different temperatures, is undoubtedly the most accurate. However, it can be applied only to substances with constant moment and accordingly, if the dipole moment of the molecule changes with temperature as in the case of 1,2-dihalogenoethanes, we have to apply another method, in which the electronic and atomic polarizations are eliminated in another way.

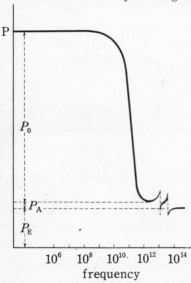

Fig. 1.14. Dependence of polarization on frequency.

If we plot the molecular polarization or the molecular refraction of a polar substance against the frequency of the radiations, over a range from long radio waves to the ultra-violet, there is obtained a curve of the form shown in Figure 1.14.

The extent of the electronic, atomic and orientation polarizations are indicated on the vertical axis. The electronic polarization, $P_E$, is effected by the field with great rapidity. The electrons have so small an inertia that the distortion of the electronic clouds keeps step with the oscillations of the external field up to very high frequencies lying far in the ultra-violet. The atomic polarization, $P_A$, is much less rapid, owing to the greater mass of the nuclei: it follows the oscillations of the field up to the infrared, but as stated above, its effect is usually small compared with $P_E$. The third term, $P_0$, is due to the orientation of the molecules that have permanent

9. P. Debye, Polar Molecules, The Chemical Catalogue Company (1929).

dipoles, and this is a relatively slow process corresponding to the period of radio waves with wave-lengths of the order of cm. $\sim$ m. (in the liquid state or in solutions).

Accordingly, if we measure the dielectric constant for waves somewhere in the visible spectrum, this will give us the electronic polarization, since there will not be time between one wave and the next for the molecules to turn round or for the nuclei within a molecule to be displaced. But as we have seen, the dielectric constant for visible light can be obtained from the refractive index by means of Eq. (1.7) which corresponds to Eq. (1.6) with the square of the refractive index in place of the dielectric constant. Thus

$$P_E = \frac{n^2 - 1}{n^2 + 2} \cdot \frac{M}{d}. \tag{1.10}$$

If we assume the atomic polarization to be small as compared with $P_E$, we may use the expression (1.10) for $P_E + P_A$ and substract this quantity from the total polarization $P$ in order to obtain $P_0$:

$$P_O = P - (P_E + P_A)$$

$$= \frac{\varepsilon - 1}{\varepsilon + 2} \cdot \frac{M}{d} - (P_E + P_A)$$

According to the Debye equation, $P_0$ is related to the dipole moment $\mu$ by:

$$P_O = \frac{4\,\pi}{3} \cdot N \cdot \frac{\mu^2}{3\,k\,T}. \tag{1.11}$$

Thus the dipole moment is calculated from the measurements of dielectric constant and refractive index at a single temperature. This is the method widely applied for the measurements of dipole moment in dilute solutions.

Evidently this is an approximate method which may introduce a serious error in the calculation of the moment value, especially that of a molecule with small moment. In order to discuss this problem in more detail, we shall again consider the change of dielectric constant with frequency. As shown by the measurements carried out by Mizushima[10] in the region of short radio waves, the dielectric constant of polar liquids decreases with decreasing wave-length of the external alternating field until it reaches a value almost equal to the square of refractive index for visible light and this region of dispersion of dielectric constant is shifted towards longer wave length as the temperature is lowered. This is explained on the basis of the Debye theory[9]

---

10. S. Mizushima, Physik. Z. **28**, 418 (1927), Sci. Pap. Inst. Phys. Chem. Res. Tokyo, **5**, 201 (1927), **9**, 209 (1928).

that the orientation of dipole molecules in an externally applied field becomes less complete at higher frequencies. Therefore, if we measure the dielectric constant of a polar liquid at frequencies sufficiently high and at temperatures sufficiently low, we can obtain a value of dielectric constant from which the value of $P_E + P_A$ can be calculated by means of Eq. (1.6). However, as a practical procedure it is more convenient to use the dielectric constant of the solid, because the region of dispersion is shifted towards longer wave lengths at which we can make measurements more conveniently.*

In the calculation of dipole moment of 1,2-dichloroethane (Section 2, Part I) we have used this kind of value for $P_E + P_A$. In this case we have spectroscopic evidence that all the molecules in the solid state are in the *trans* form with no moment, and accordingly the molecular polarization calculated from the dielectric constant of the solid, should be exactly equal to $P_E + P_A$ of the *trans* molecule, irrespective of the frequency of the external field. The corresponding quantity of the *gauche* molecule may be a little different from this value and even the atomic polarization of the *trans* molecule in the solid state may be different from that in the liquid and gaseous states. Actually if there is a difference, it should be explained in terms of the difference in intramolecular motion between the solid state and the liquid or gaseous state. It will easily be seen that the most probable difference is found in the torsional motion about the C–C axis and this motion in the liquid and the gaseous states will take place with larger amplitude than in the solid state. If this torsional motion has larger amplitude, it will make more contribution to the atomic polarization.

It will not be out of place to explain here in more detail the relation between the atomic polarization and the infrared absorption, since this problem bears an important relation to that of internal rotation as well as to infrared spectroscopy.

If the vibration of the molecule is harmonic, the atomic polarization will be expressed as

$$P_A = \frac{N}{9\,\pi} \sum_i \frac{e_i{}^2}{m_i\,(v_i{}^2 - v_0{}^2)}, \qquad (1.12)$$

where $e_i$ is the effective charge, $m_i$ the reduced mass, $v_i$ the vibrational frequency of the molecule, $v_0$ the frequency of the incident wave, and $N$

---

* See, for example, C. P. Smyth, Dielectric constant and molecular structure, Chemical Catalogue Company, New York (1931) and reference 9.

the Avogadro number.[11] Usually the dielectric measurement is made at very low frequency or in the static field, and the foregoing expression can be reduced to:

$$P_A = \frac{N}{9\,\pi} \sum_i \frac{e_i{}^2}{m_i\,v_i{}^2}.$$ (1.13)

Therefore, if we know the effective mass and the vibrational frequencies which can be observed as infrared absorption, we can calculate the value of the atomic polarization. The effective charge which appears in this expression need not be at all like the charge of an electron or nucleus, as it is by definition not the charge of a single particle, but rather the differential coefficient $d\mu/dr$ of the dipole moment with respect to nuclear distance. For instance, the effective charge of a nonpolar diatomic molecule, such as $H_2$, $N_2$, etc., is zero and accordingly the atomic polarization of such a molecule is zero irrespective of the value of the vibrational frequency. (Such a vibration does not appear as the infrared absorption.)

It will be seen from Eq. (1.13) that a normal vibration with lower frequency will make a greater contribution to the atomic polarization, if the other conditions are kept constant. Accordingly, it would be of interest to calculate the value of atomic polarization by taking into account the lower normal frequencies. Such calculations have been made in our laboratory and have yielded some interesting results.[12]

In Section 14, Part I, a question has been raised about the explanation of the moment value of 1,4-dioxane which affects the conclusion about the stable forms of this molecule. If the value (0.4D) actually arises from the orientation polarization, we have to conclude that some of the molecules exist in boat form, while if this value is attributable to atomic polarization, almost all the molecules will exist in the nonpolar chair form. According to the calculation just referred to, the value seems to be explained as atomic polarization. Another interesting example refers to the reported large moment (1.3D) of 1,4-cyclohexanedione.[13] This would require the existence of a considerable amount of the polar form in addition to the nonpolar chair form:

11.  J. H. Van Vleck, The Theory of Electric and Magnetic Susceptibilities, Oxford at the Clarendon Press (1932).
12.  T. Chiba, I. Miyagawa, M. Yasumi and M. Shirai, private communication.
13.  C. G. Le Fèvre and R. J. W. Le Fèvre, J. Chem. Soc., 1935, 1696.

$$\begin{array}{c} \qquad\qquad C \diagup\!\!\!\!^{\displaystyle O} \\[2pt] \left|\!\!\begin{array}{c} CH_2\text{-}CH_2 \diagup \\ \diagdown CH_2\text{-}CH_2 \end{array}\!\!\right| \\[2pt] O \diagup\!\!\!\!^{\displaystyle C} \end{array}$$

However, some deformation vibration of this chair form will give rise to a strong absorption in the low frequency region due to the large bond moment of the $C = O$ group. This would make a large contribution to the atomic polarization, which would explain reasonably the reported large moment, even if we assume that the majority of molecules are in the nonpolar form.

Yasumi, Shirai and Mizushima[14] found a large value of atomic polarization (about 14 c. c.) for several monovalent alcohols in the liquid state: this value is almost seven times as large as that found in the gaseous state. This seems to be due to the intermolecular hydrogen bond in the liquid state, since the value of atomic polarization is independent of the length of aliphatic chain. Presumably this is due to vibration or rotatory vibration of bonded OH groups which gives rise to a strong infrared absorption.

As stated in the beginning of this section, the methods employed for the determination of dipole moments depend almost wholly on the measurement of dielectric constant. The experimental method most commonly employed is the heterodyne beat method. Two similar electron-tube oscillators, containing suitable capacities and inductances to give a frequency of about $10^6$ cycles per sec. are coupled to an amplifier. One of the oscillators contains fixed condensers and hence has a fixed frequency, whereas the other has an accurately calibrated standard variable condenser in parallel with the experimental cell: the latter may consist of two or more concentric metal, or metal-coated, cylinders capable of acting as a condenser. By adjustment of the variable condenser the frequency of the oscillator can be altered so as to coincide with that in the fixed oscillator: no sound can then be detected in the amplifier. More accurate measurement can be made by listening for a beat note of definite frequency, which is equal to the frequency difference between the two oscillators. If the readings of the standard condenser are taken, with the experimental cell in and out of the circuit, the capacity can be determined, both empty and full of the substance to be examined: the

---

14. M. Yasumi, M. Shirai and S. Mizushima, Bull. Chem. Soc. Japan, **25**, **133** (1952).

dielectric constant can then be evaluated.  In the resonance method, which is also often employed, a fixed frequency oscillator is coupled to a resonance circuit containing inductance and capacity, the latter in the form of a standard variable condenser and the experimental cell.  At a given frequency, determined by the inductance and capacity, resonance occurs between the two systems, and this is indicated by the maximum deflection of a galvanometer associated with the resonance circuit.  The readings of the standard condenser are then taken, with and without the experimental cell, and the dielectric constant calculated.  In practice various corrections must be applied with both methods, e. g., for the capacity of leads: this is generally done by making measurements on a standard substance of known dielectric constant, e. g., air or carbon dioxide for gases, and benzene for liquids.

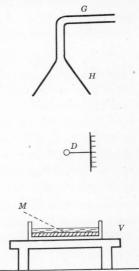

A detailed explanation of the experimental methods outlined above is not ⊁ necessary, since they are usual methods for the measurement of dielectric constant.  In our laboratory the heterodyne beat method[15] is used for the determination of dipole moment and the resonance method[16] is applied for the measurement of dispersion of dielectric constant.  As to the dispersion measurement, it often becomes necessary to work at very short wave lengths at which special experimental techniques are required. As one of such techniques we shall explain, in the following, the free wave method which has been developed in our laboratory.[17]

Fig. 1.15. Measurement of refractive index and absorption coefficient by free wave method.

In this method we measure the reflection coefficient of a dielectric sheet placed on a metallic plate.  A microwave of constant wave length is emitted vertically into the free space from a horn $H$ through a wave guide $G$. (Fig. 1.15).  A wooden vessel $V$ with an area sufficient to remove the edge

15.  S. Mizushima, Y. Morino and K. Higasi, Sci. Pap. Inst. Phys. Chem. Res. Tokyo, **25**, 159 (1934).

16.  S. Mizushima, Sci. Pap. Inst. Phys. Chem. Res. Tokyo, **5**, 201 (1927).

17.  M. Yasumi, K. Nukazawa and S. Mizushima, Bull. Chem. Soc. Japan, **24**, (1951).

effect is placed horizontally under $H$. A crystal detector $D$ (so adjusted that the deflection of the galvanometer is proportional to the energy of the wave) is moved vertically between $H$ and $V$ and the energy distribution of the standing wave is determined.

The experimental procedure is as follows: At first the mercury is poured into the vessel $V$ to form a horizontal metallic surface $M$. A standing wave is formed between $H$ and $M$. If the amplitude of the incident wave is $E_0$, the maximum and the minimum energies of the standing wave will be proportional to $4\,E_0{}^2$ and zero, respectively, provided that there is no reflection of wave other than that from the surface of the mercury.

Next, we place on the mercury the liquid layer of the dielectric whose thickness can be calculated accurately from the volume of the liquid and the inner area of the vessel. If the liquid absorbs the microwave, the incident and the reflected wave will form an imperfect standing wave, and the maximum and the minimum energies will become proportional to $E_0{}^2\,(1+R)^2$ and $E_0{}^2\,(1-R)^2$, respectively, where $R$ is the reflection coefficient. In other words, the ratio of the extreme values will be $(1-R)^2/(1+R)^2$. This ratio can be determined experimentally by moving the detector $D$ vertically and reading the galvanometer deflections. The value of $R$ can at once be calculated from this ratio. By making such measurements at different thicknesses $d$ of the dielectric layer, we obtain a relation between $R^2$ and $d$, from which we can calculate the refractive index $n$ and absorption coefficient $k$ by use of the following relation.

$$R^2 = \frac{\sinh^2\left(\dfrac{2\pi k d}{\lambda} + \dfrac{1}{2}\ln R_{12}\right) + \cos^2\left(\dfrac{2\pi n d}{\lambda} - \dfrac{\gamma}{2}\right)}{\sinh^2\left(\dfrac{2\pi k d}{\lambda} - \dfrac{1}{2}\ln R_{12}\right) + \cos^2\left(\dfrac{2\pi n d}{\lambda} + \dfrac{\gamma}{2}\right)} \qquad (1.14)*$$

where

$$R_{12}{}^2 = \frac{(n-1)^2 + k^2}{(n+1)^2 + k^2}$$

$$\tan\gamma = \frac{2\,k}{(n^2-1) + k^2}.$$

---

* Yasumi derived a general formula for the reflection by multiple layer of dielectrics. See M. Yasumi, Bull. Chem. Soc. Japan, 24, 53 (1951).

## 4. Electron Diffraction

The first experiment of the diffraction of fast electrons by gases was made by Mark and Wierl[18] a few years after the experimental proof by Davisson and Germer of the wave nature of electron beams. According to de Broglie a beam of electrons with velocity $v$ is associated with a wave train of the wave length $\lambda = h/m\,v$, where $h$ is Planck's constant and $m$ the mass of the electron. If $V$ is the voltage applied between the cathode and anode to produce an electron beam of velocity $v$, the wave length is calculated as:

$$\lambda = \frac{h}{m\,v} = \sqrt{\frac{150.5}{V}} \text{ Å,} \qquad (1.15)$$

since

$$\frac{1}{2}\,m\,v^2 = e\,V.$$

Thus, for $V = 45000$ volts the wave length becomes 0.058Å, which is twenty times as short as the wave length of X-rays ordinarily used in diffraction experiments. If such an electron beam falls on a molecule consisting of a number of atoms, the secondary beams scattered by the molecule interfere with one another, and produce diffraction halos. If the applied voltage is higher than 1000 volts, the relativity correction for the wave length must be taken into consideration. The accurate expression is given by

$$\lambda = \sqrt{\frac{150.5}{V}}\,(1 - 4.9 \times 10^{-7}\,V) \text{ Å.} \qquad (1.16)$$

However, the wave length of the electron beam is usually measured at each exposure, using gold foil as a standard without making such a calculation.

For gas molecules orientated at random, the intensity of the scattered electrons is given by the expression:

$$I = \frac{I_0}{R^2}\,\frac{64\,\pi^4\,m^2\,e^4}{h^4}\,\frac{1}{s^4}\left[\sum_i \sum_j (Z_i - F_i)\,(Z_j - F_j)\,e^{-\alpha_{ij}\,s^2}\,\frac{\sin s\,r_{ij}}{s\,r_{ij}}\right.$$

$$\left. + \sum_i (Z_i - F_i)^2 + \sum_i (Z_i - S_i)\right] \qquad (1.17)$$

where

$$s = \frac{4\,\pi}{\lambda}\,\sin\frac{\theta}{2}.$$

18. H. Mark and R. Wierl, Naturwissenschaften, 18, 205 (1930).

Here $I_0$ is the intensity of the incident beam, $R$ the distance from the scattering molecule to the point of observation, $r_{ij}$ the distance between the $i$-th and $j$-th atoms, $\theta$ the angle of scattering, $Z_i$ the atomic number, and $F_i$ the atom form factor for X-rays. In the derivation of this equation each atom of the molecule is considered to have an electron cloud spherically symmetrical about the nucleus. $\alpha_{ij}$ is related to the mean amplitudes $\langle \Delta\, r_{ij}^2 \rangle$ of the thermal vibration of atoms by the expression:

$$\alpha_{ij} = \frac{1}{2}\langle \Delta\, r_i^2 \rangle. \qquad (1.18)$$

The first term of Eq. (1.17) is called "molecular scattering term", and is the part in which we are interested for the determination of molecular structure. The second term denotes the coherent scattering by single atoms, and the third term the incoherent scattering due to the inelastic collision of electrons with the atoms. The last two terms produce only the background decreasing monotonously with increasing angle.

The apparatus used in our laboratory was constructed by Morino according to the advice of Prof. R. Ueda of Nagoya University. It is of usual type, as shown in Fig. 1.16 except that the gas nozzle is drum-shaped. The gas to be examined is introduced through the side tube and is spread into the volume of the apparatus from the center hole of the drum. On leaving this hole the gas molecules cross the electron beam, and diffract it. The diffraction halos are photographed on a plate at a distance of 110 mm. from the nozzle.

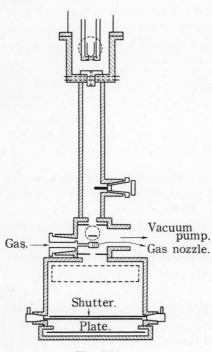

Fig. 1.16.

Electron diffraction apparatus for gas.

*A. Visual method.* Equation (1.17) indicates that the intensity of scattered beam is related to molecular structure through atomic distances $r_{ij}$. The fourth power term in the denominator causes such a rapid decrease in intensity with increasing angle that the maxima and minima due to the molecular term are, as a rule, undetectable by the microphotometer recording. Nevertheless, the eye apparently discounts the steady decrease in the background and interprets any change in the rate of decrease as a maximum or a minimum. Also the eye detects many rings far out from the central spot which do not show up in the microphotometer records.

On this account the visual method was suggested by Pauling and Brockway[19] and revised by Schomaker[20]. The standard procedure of our laboratory (Morino et al.) is as follows. First the points of apparent maxima and minima of halos are marked on the plate by simple observation by eyes, the diameters of halos are measured referring to these points, and then the visual curve is drawn by estimating the relative intensities and the shapes of the halos. This is compared with the theoretical intensity curves calculated for a number of plausible models, using a simplified equation:

$$I = K \sum_i \sum_j Z_i Z_j e^{-\alpha_{ij} s^2} \frac{\sin s\, r_{ij}}{s\, r_{ij}} \qquad (1.19)$$

which is identical with the equation used in Section 8, Part I except for the factor $e^{-\alpha_{ij} s^2}$ of the mean amplitude of the thermal vibration of atoms. In this equation the atom form factor $F$ is omitted, since it decreases rapidly with increasing angle and it can be neglected at the angles where the comparison is actually made. Moreover, the term $s^4$ is neglected, because it is the position of a halo, and not the intensity, that we can accurately determine by the visual method. By the comparison of the theoretical and visual curves the best molecular model is selected by trial and error method.

B. *Radial distribution method.* The disadvantage of the method just described is that it is not a straightforward process for determining the structure of a molecule from the experimental results. It is usually unavoidable to calculate many theoretical intensity curves before the best fit is obtained, and, moreover, the final result obtained cannot be considered to be the only one fitting the experimental result. Hence, it is desirable to devise a direct method of deducing interatomic distances from the diffraction photographs.

---

19. L. Pauling and L. O. Brockway, J. Chem. Phys., **2**, 867 (1934).

20. V. Schomaker, quoted by H. D. Springall and L. O. Brockway, J. Am. Chem. Soc., **60**, 996 (1938).

Pauling and Brockway have introduced a radial distribution method for this purpose.[21] Putting aside the atomic scattering terms, we shall consider only the molecular scattering term in the form of Eq. (1.19). If we arrange all the atom pairs in the order of their distances $r_{ij}$ with their values of $Z_i Z_j$, we get a kind of density function. Let this function be expressed as $r D(r)$, where $D(r)$ is the radial distribution function.[22] Then $D(r)$ is equal to $Z_i Z_j / r_{ij}$ wherever $r$ is equal to $r_{ij}$ and is equal to zero otherwise, and accordingly, it is a discontinuous function. Actually the thermal vibration gives a finite density in the vicinity of the equilibrium distance $r_{ij}$, which can be approximated by an error function about the equilibrium distance. Therefore, $D(r)$ is really a continuous function of $r$. Using this function in place of the factor $Z_i Z_j / r_{ij}$, the double sum of Eq. (1.19) can be replaced by an integral:

$$s I(s) = K \int_0^\infty D(r) \sin s\, r\, dr \tag{1.20}$$

This expression is the Fourier integral for $s I(s)$, the coefficients of the Fourier terms being $D(r)$. On inverting the Fourier integral we obtain the following expression for $D(r)$ in terms of $I(s)$:

$$D(r) = K' \int_0^\infty s I(s) \sin s\, r\, ds. \tag{1.21}$$

Instead of this Pauling and Brockway used an expression in which the integral is replaced by a finite sum of terms, one for each maximum or minimum. Recently, however, Eq. (1.21) has been used in its original form, yielding a better radial distribution curve. This is due on the one hand to the simplification of calculation by the application of punched card methods and on the other hand to the acquisition of the accurate intensity curve by the sector method.

In an actual case the intensity is measured only in a finite region, and the integration of Eq. (1.21) cannot be performed to infinity. In consequence there is a possibility of the appearance of a ghost maximum by "diffraction effect". In order to avoid this, a factor $e^{-\alpha s^2}$ was inserted dy Degard,[23]

---

21. L. Pauling and L. O. Brockway, J. Am. Chem. Soc., **57**, 2684 (1935).

22. I. L. Karle and J. Karle, J. Chem. Phys., **17**, 1052 (1949). These authors defined $D(r)$ as shown above. Usually the quantity referred to above is expressed as $r^2 D'(r)$ and $D'(r)$ is called radial distribution function.

23. C. Degard, Bull. Soc. Roy. Sci. Liege **12**, 383 (1937).

in the integrand of Eq. (1.21), where $\alpha$ is a constant whose value is chosen so that

$$e^{-\alpha s^2} = 0.1$$

at the outermost end of the observed values.  Thus the modified radial distribution function is expressed as:

$$f(r) = K'' \int_0^{s_{max}} s\, I(s)\, e^{-\alpha s^2} \sin s\, r\, ds,\qquad (1.22)$$

where $s_{max}$ is the range of scattering data.  It will easily be seen that the introduction of this factor does not change the position of a maximum, but makes every maximum flatter, and transforms the sharp peak of $D(r)$ into a bell-shaped curve with the breadth of $\alpha^2/2$.

Mathematically the radial distribution method is almost equivalent to the visual method explained above.  However, in nearly all the practical cases both of them are used together because of the directness of the radial distribution method and the greater sensitivity of the intensity curve to some configurational features.

C. *Punched card method*.  The calculation of the intensity function and radial distribution function can be reduced to the evaluation of the expression:

$$F(x) = \sum_i G_i \sin \frac{\pi}{10} y_i\, x,\qquad (1.23)$$

where in the case of the simpler molecules we have twenty or more terms for the intensity function and about 100 terms for the radial distribution function.  At first such calculations were made with the aid of Sherman's table and an adding machine.  In the punched card method the values of $G_i \sin y_i\, x$ are punched on a set of cards for certain values of $G_i$ and for a series of values of $y_i$ and $x$.  These values are so chosen that any required value of $G_i \sin y_i\, x$ can be obtained by taking one or several cards.  The summation of these values is made by the use of the tabulating machine.

Some investigators used 300,000  cards, while Morino cut down the number of cards to only 60,000 without sacrificing the accuracy.  He calculated the values of $G_i \sin y_i\, x$ in four figures, so that twenty values of $G_i \sin y_i\, x$ could be punched on a card of the Remington-Rand machine.  He prepared two sets of cards, one for the intensity curve, and the other for the radial distribution function.  The former contains 25,000 cards for $y = 0.01 \sim 5.00$ $(\Delta y = 0.01)$, $x = 0 \sim 99$ $(\Delta x = 1)$, $G = 1, 5, 10, 50$, the latter 10,000 cards for $y = 1 \sim 100$ $(\Delta y = 1)$, $x = 0.05 \sim 10.00$ $(\Delta x = 0.05)$,

$G = 1, 5, 10, 50$. In addition to these cards 14,000 mother cards and 10,50 index cards are prepared. The use of punched cards enables us not onl to save time but also to make more complicated calculations.

*D. Sector method.* As stated above the molecular scattering term is maske considerably by the background falling rapidly with increasing angle, s that only humps are observed in the microphotometer records. Finbak[24] and later P. P. Debye[25] proposed to use in front of the photographic plat a rotating sector producing a screening effect, large for smaller scatterin angle and rapidly decreasing with increasing angle. The photometer record of such plates are expected to give us more precise informations concernin the molecular scattering curve. Finbak and Hassel constructed a convenien apparatus of this type. With this kind of apparatus Karle and Karle[22] wer able to measure even the mean amplitudes of thermal agitation in atomi distances in a molecule in addition to the accurate values of equilibriun distances. Recently T. Ino[26] constructed a simple apparatus with rotatin sector which has been found to be very useful for the structure investigation

In conclusion, the visual method can be applied with good results to th determination of atomic distances in a molecule if the diffraction patter consists of well defined maxima and minima. If, however, the structur investigation requires the determination of minute features of the pattern such as shelves or asymmetries, the visual method may be less reliable In such a case it is desirable to apply the sector method with the radia distribution method.

We should like to add only a few lines on the mean amplitude of therma vibrations of atom pairs referred to above. This problem has been discusse by Morino et al.[27, 28] who derived the mean square amplitude as:

$$\langle \Delta R_i^2 \rangle = k\,T\,F_{ii}^{-1} + \frac{h^2}{64\,\pi^2\,k\,T}\,(\mu + \mu')$$

where $\mu$ and $\mu'$ are the reciprocals of the masses of the two atoms formin the bond under consideration and $F^{-1}$ is the inverse matrix of $F$ (or th potential energy matrix of the vibrational problem) of which we shall explai in detail in the next chapter. In a similar way they calculated the mea square amplitude for atom pairs not bonded directly.

24. C. Finbak, Avhand. Norske Vid.-Akad. Oslo, Mat.-Naturv. Kl., No. 13 (1937)
25. P. P. Debye, Physik. Zeits., 40, 66, 404 (1939).
26. T. Ino, J. Phys. Soc. Japan, 8, 92 (1953).
27. Y. Morino, K. Kuchitsu and T. Shimanouchi, J. Chem. Phys., 20, 726 (1952)
28. Y. Morino, Private communication.

# CHAPTER II

# Normal Vibrations

## 5. Normal Vibrations

The mechanical motions of a molecule containing more than two atoms are somewhat more complicated than those of a diatomic molecule and hence the resulting spectrum is quite complex. If the molecule contains $N$ atoms, $3N$ coordinates are required to describe its motion. Of these $3N$, three correspond to translation of the molecule as a whole. These are of little interest to the spectroscopist. Three more coordinates belong to the rotation of the molecule as a whole. These account for the rotational structure of the Raman and infrared bands. The remaining $3N - 6$ coordinates determine the vibrational motion of the molecule, unless the molecule is linear, when there are $3N - 5$ vibrations.

Classical mechanics may be used to a large extent in discussing the motion and, as a first approximation, all of the nuclei are fixed so that the molecule cannot rotate. The kinetic and potential energies of the molecule are now expressed in terms of some appropriate coordinates, $q_i$ and the corresponding velocities, $\dot{q}_i$. The kinetic energy, $T$ becomes:

$$2\,T = \sum a_{ij}\,\dot{q}_i\,\dot{q}_j.$$

where the $a_{ij}$ are some functions of the masses of the nuclei, and if the motion is simple harmonic the potential energy, $V$ is given ·by:

$$2\,V = \sum b_{ij}\,q_i\,q_j,$$

where the $b_{ij}$ are force constants. The appearance of cross product terms in $T$ and $V$ may be overcome by a linear transformation to new coordinates:

$$q_k = \sum B_{kl}\,Q_l,$$

after which $T$ and $V$ become

$$2\,T = \sum \dot{Q}_i{}^2,$$

$$2\,V = \sum \lambda_i\,Q_i{}^2$$

185

and then the equations of motion are easily solvable. We have

$$\ddot{Q}_k + \lambda_k Q_k = 0$$

The new coordinates, $Q_k$ are called normal coordinates and the solution of the equations of motion shows that each nucleus executes a simple harmonic oscillation about its equilibrium position with frequency $\nu_k$,

$$2\pi \nu_k = (\lambda_k)^{1/2}.$$

All of the nuclei move with the same frequency and the same phase: that is, they all pass through their equilibrium positions and reach their positions of maximum amplitude at the same time. The complete motion of the molecule is thus complex but it may be considered as a superposition of $3N - 6$ normal vibrations, each with its own frequency.

This rather abstract description of normal vibrations may be visualized by analogy with a mechanical model. Suppose that a model of the carbon dioxide molecule, $CO_2$, is constructed by using weights in the ratio of 12 to 16, respectively, for the carbon and oxygen atoms, these weights being held in proper orientation by suitable springs. Suppose further that the carbon oxygen springs are now stretched in such a way that two oxygen atoms are moved away symmetrically from the central carbon atom as shown in Fig. 2.1 (a). If, now, the weights are released simultaneously, a vibration will occur in which the weights move back and forth along the connecting bonds. This is a characteristic vibration of the model, inasmuch as it has a definite frequency which can be measured by a stroboscope, and it does not excite any other vibration in the model. This stretching motion is exactly analogous to one of the $3N - 5$, or 4 normal modes of the carbon dioxide molecule. Another carbon dioxide vibration can be visualized readily by pulling the central carbon weights above the axis of the model while the oxygen weights are pulled down as shown in Fig. 2.1 (c) and then releasing all three weights simultaneously. This vibration is doubly degenerate and the remaining vibration has a mode as shown in Fig. 2.1 (b).

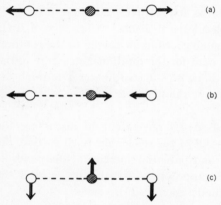

(a)

(b)

(c)

Fig. 2.1 Vibrational modes of $CO_2$.

Now suppose that the model is struck a blow with a hammer. The weights will perform in general a complicated motion having no apparent relation to the individual modes referred to above. However, if this apparently random motion is photographed with a stroboscopic camera adjusted successively for each of the frequencies of the normal modes, each of these modes of vibration will be found to be faithfully performed by the weights.

An infrared spectrometer plays the same role with respect to the actual carbon dioxide molecule as the stroboscope does to the model. Thus, by measuring the frequencies of the infrared radiation absorbed by a substance the spectrometer determines the characteristic mechanical frequencies of its molecule.

A question arising at this point concerns the practical mathematical calculation of the normal modes of vibration. Such calculations we shall make in the following sections for the 1,2-dihalogenoethanes and the $n$-paraffins, but we should like to point out at this stage that for more complex molecules, a rigorous treatment becomes impossible. Accordingly some other method is required to correlate the observed spectral frequencies with the structure of the molecule. Considerable success in this direction has been achieved by a purely empirical approach.

In order to explain the basis for such an empirical method, let us consider, for example, a mechanical model of a molecule containing only one C–H bond, such as pentachloroethane referred to in Part I. If the C–H spring is stretched and released, the hydrogen weight vibrates rapidly against the carbon with a characteristic frequency. The other weights, on the other hand, are so heavy that they are almost totally unable to follow the vibration. It is true, at least to a first approximation, that the observed stretching frequency is a characteristic of the C–H spring (bond) and the mass of hydrogen atom, and is practically independent of the rest of the molecule. Similarly, a bending or deformation can be studied by displacing the hydrogen weight in a direction normal to the C–H direction and then releasing it. Again the hydrogen weight will move characteristically with the remaining weights practically at rest.

Thus we may consider characteristic C–H frequencies which are almost independent of the structure of the rest of the molecule. Actually a study of hundreds of molecules containing C–H linkages has shown an absorption band or Raman line around 2900 cm.$^{-1}$ (C–H stretching) and another around 1450 cm.$^{-1}$ (C–H bending). Further verification may be found by a study of the absorption spectra of molecules in which the hydrogen atom has been replaced by a deuterium atom of mass two where we have approximately the relation:

$$\sqrt{2}\,\nu_{\text{C-D}} = \nu_{\text{C-H}}.$$

If, therefore, we compare the spectra of a large number of different molecules having a common atomic group, we may find an absorption band or a Raman line whose frequency remains constant throughout the series. As such an example we have often referred to the N–H frequency in Part I.

It is, however, evident that it is not possible to ascribe every observed absorption band or Raman line to a specific atomic group. Indeed, if this were true, it would make difficult the possibility of differentiating between isomeric compounds just as we have done in the case of rotational isomers of the 1,2-dihalogenoethanes. In that case we have used a term such as "C–Cl frequency" only to emphasize that this bond has the large amplitude in this vibration, but it is evident that this frequency arises from a normal mode of vibration which is characteristic of the molecule as a whole. We shall discuss this problem in more detail in Section 9.

## 6. Normal Vibrations of the 1,2-Dihalogenoethanes

In the preceding section we have dealt with the problem of normal vibrations rather formally and it is necessary to show at least an example by which a beginner in this field can understand an actual process of calculation. As an example, we shall make, in the following paragraphs, the normal vibration calculation of the skeletons of the 1,2-dihalogenoethanes X–H$_2$C–CH$_2$–X, the result of which we have often used in the explanation of structural problems in Part I.

The calculation of skeletal vibrations of the 1,2-dihalogenoethanes was first made by Mizushima and Morino,[1] using a valence force field. Afterwards Mizushima, Morino and Shimanouchi[2] improved this calculation, using the Urey-Bradley field which had been shown by Shimanouchi[3] to be adequate to explain the vibrational spectra of many simple molecules.

The potential energy change due to a change in bond length $(\Delta r_i)$ and bond angle $(\Delta \alpha_i)$ can be expressed as:

$$V = \frac{1}{2}\sum_i K_i\,(\Delta r_i)^2 + \frac{1}{2}\sum_i H_i\,(\Delta \alpha_i)^2 + \frac{1}{2}\sum_i F_i\,(\Delta q_i)^2 \qquad (2.1)$$

1.  S. Mizushima and Y. Morino, Sci. Pap. Inst. Phys. Chem. Res. Tokyo, 26, 1 (1934).
2.  S. Mizushima, Y. Morino, and T. Shimanouchi, Sci. Pap. Inst. Phys. Chem. Res. Tokyo, 40, 87 (1942).
3.  T. Shimanouchi, J. Chem. Phys., 17, 245, 734, 848 (1949).

where $\Delta q_i$ denotes the change in distance between non-bonded atoms and $K_i$, $H_i$, $F_i$ the force constants. However, $\Delta r_i$, $\Delta \alpha_i$, and $\Delta q_i$ in Eq. (2.1), are not independent of one another and accordingly in order to express the potential field more accurately it is necessary to take into account the first order terms with respect to $\Delta r$, $\Delta \alpha_i$ and $\Delta q_i$.

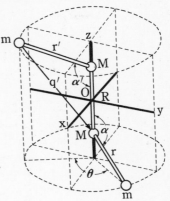

$$V = \sum_i \left[ K_i{}' r_{i0} (\Delta r_i) + \frac{1}{2} K_i (\Delta r_i)^2 \right]$$

$$+ \sum_i \left[ H_i{}' r_{i\alpha}{}^2 (\Delta \alpha_i) + \frac{1}{2} H_i r_{i\alpha}{}^2 (\Delta \alpha_i)^2 \right]$$

$$+ \sum_i \left[ F_i{}' q_{i0} (\Delta q_i) + \frac{1}{2} F_i (\Delta q_i)^2 \right] \qquad (2.2)$$

Fig. 2.2.
Skeleton of X–$H_2$C–$CH_2$–X.

Here $r_{i0}$, $r_{i\alpha}$ and $q_{i0}$ are the equilibrium values of distances which are inserted in order to make all the force constants, $K_i{}'$, $H_i{}'$, $F_i{}'$, $K_i$, $H_i$ and $F_i$ dimensionally similar.

Figure 2.2 shows the configuration of the skeleton of $XH_2C$–$CH_2X$. The notations used in this figure as well as in the mathematical expressions have the following significance,

$x_1, y_1, z_1$: Cartesian coordinates of one halogen atom X,
$x_2, y_2, z_2$: those of one $CH_2$ group,
$x_3, y_3, z_3$: those of the other $CH_2$ group,
$x_4, y_4, z_4$: those of the other halogen atom X,

$m$:  mass of halogen atom,
$M$:  mass of $CH_2$ group,
$r, r'$:  the C–X bond lengths,
$R$:  the C–C bond length,
$\alpha, \alpha'$:  the C–C–X bond angles,
$q, q'$:  distances between C and X not bonded directly,
$\theta$:  azimuthal angle of internal rotation about C–C axis ($\theta = 180°$ corresponding to the *trans* form and $\theta = 60°$ to the *gauche* form),
$r_0$:  equilibrium value of $r$ and $r'$,
$R_0$:  equilibrium value of $R$,
$\alpha_0$:  equilibrium value of $\alpha$ and $\alpha'$,
$q_0$:  equilibrium value of $q$ and $q'$,

$$\mu = 1/m, \quad \mu_0 = 1/M,$$
$$\rho_0 = (R_0/r_0)^{\frac{1}{2}}, \quad \rho = (r_0/R_0)^{\frac{1}{2}},$$
$$r_\alpha = (r_0 R_0)^{\frac{1}{2}}.$$

In terms of these notations Eq. (2.2) can be rewritten as Eq. (2.3), where we neglect the interaction between the two halogen atoms which are far apart from each other.

$$V = K_R' R_0 (\varDelta R) + \frac{1}{2} K_R (\varDelta R)^2 + K_r' r_0 (\varDelta r) + \frac{1}{2} K_r (\varDelta r)^2$$

$$+ K_r' r_0 (\varDelta r') + \frac{1}{2} K_r (\varDelta r')^2 + H' r_a (r_a \varDelta \alpha) + \frac{1}{2} H (r_\alpha \varDelta \alpha)^2$$

$$+ H' r_a (r_a \varDelta \alpha') + \frac{1}{2} H (r_a \varDelta \alpha')^2 + F' q_0 (\varDelta q) + \frac{1}{2} F (\varDelta q)^2 \tag{2.3}$$

$$+ F' q_0 (\varDelta q') + \frac{1}{2} F (\varDelta q')^2.$$

Here the significance of the force constants $K_R'$, $K_r'$, $H'$, $F'$, $K_R$, $K_r$, $H$ and $F$ will be clear.

It is readily seen from Fig. 2.2:

$$q^2 = R^2 + r^2 - 2 R r \cos \alpha$$
$$q'^2 = R^2 + r'^2 - 2 R r' \cos \alpha'$$

hence

$$\varDelta q = s_0 (\varDelta R) + s_1 (\varDelta r) + \sqrt{t_0 t_1} (r_a \varDelta \alpha) + \frac{1}{2 q_0} [t_0^2 (\varDelta R)^2 + t_1^2 (\varDelta r)^2$$

$$- s_0 s_1 (r_a \varDelta \alpha)^2 - 2 t_0 t_1 (\varDelta R) (\varDelta r) + 2 t_0 s_1 (\varDelta R) (r_0 \varDelta \alpha) + 2 t_1 s_0 (\varDelta r) (R_0 \varDelta \alpha)]$$

where

$$s_0 = (R_0 - r_0 \cos \alpha_0)/q_0$$
$$s_1 = (r_0 - R_0 \cos \alpha_0)/q_0$$
$$t_0 = r_0 \sin \alpha_0/q_0 \tag{2.4}$$
$$t_1 = R_0 \sin \alpha_0/q_0$$

Similar expressions are obtained for $q'$. With these relations we can rewrite Eq. (2.3) as

$$V = (K_R{}' R_0 + 2 F' q_0 s_0)\,(\varDelta R) + (K_r{}' r_0 + F' q_0 s_1)\,[(\varDelta r) + (\varDelta r')]$$

$$+ [(H' r_a + F' q_0 (t_0 t_1)^{1/2})]\,[(r_a \varDelta \alpha) + (r_a \varDelta \alpha')] + \frac{1}{2}\,(K_R + 2 t_0{}^2 F' + 2 s_0{}^2 F)\,(\varDelta R)^2$$

$$+ \frac{1}{2}\,(K_r + t_1{}^2 F' + s_1{}^2 F)\,[(\varDelta r)^2 + (\varDelta r')^2] \tag{2.5}$$

$$+ \frac{1}{2}\,(H - s_0 s_1 F' + t_0 t_1 F)\,[(r_a \varDelta \alpha)^2 + (r_a \varDelta \alpha')^2]$$

$$+ (- t_0 t_1 F' + s_0 s_1 F)\,[(\varDelta R)(\varDelta r) + (\varDelta R)(\varDelta r')]$$
$$+ \rho\,(t_0 s_1 F' + s_0 t_1 F)\,[(\varDelta R)(r_a \varDelta \alpha) + (\varDelta R)(r_a \varDelta \alpha')]$$
$$+ \rho_0\,(t_1 s_0 F' + s_1 t_0 F)\,[(\varDelta r)(r_a \varDelta \alpha) + (\varDelta r')(r_a \varDelta \alpha')].$$

In this equation all the variables, $\varDelta R, \varDelta r, \varDelta r', \varDelta \alpha$ and $\varDelta \alpha'$ are independent of one another and evidently the linear terms or the first three terms vanish. Now let us rewrite Eq. (2.5) as follows:

$$V = \frac{1}{2} \sum_{kl} F_{kl} R_k R_l \qquad (k, l = 1, 2, 3, 4, 5) \tag{2.6}$$

where

$$\begin{aligned} R_1 &= \varDelta R \\ R_2 &= \varDelta r \\ R_3 &= \varDelta r' \\ R_4 &= r_\alpha \varDelta \alpha \\ R_5 &= r_\alpha \varDelta \alpha' \end{aligned} \tag{2.7}$$

and $F_{kl}$ is the element of matrix $F$:

$$F = \left\{ \begin{array}{lll} K_R + 2 t_0{}^2 F' + 2 s_0{}^2 F & -t_0 t_1 F' + s_0 s_1 F & -t_0 t_1 F' + s_0 s_1 F \\ -t_0 t_1 F' + s_0 s_1 F & K_r + t_1{}^2 F' + s_1{}^2 F & 0 \\ -t_0 t_1 F' + s_0 s_1 F & 0 & K_r + t_1{}^2 F' + s_1{}^2 F \\ \rho\,(t_0 s_1 F' + s_0 t_1 F) & \rho_0\,(t_1 s_0 F' + s_1 t_0 F) & 0 \\ \rho\,(t_0 s_1 F' + s_0 t_1 F) & 0 & \rho_0\,(t_1 s_0 F' + s_1 t_0 F) \end{array} \right.$$

$$\left. \begin{array}{ll} \rho\,(t_0 s_1 F' + s_0 t_1 F) & \rho\,(t_0 s_1 F' + s_0 t_1 F) \\ \rho_0\,(t_1 s_0 F' + s_1 t_0 F) & 0 \\ 0 & \rho_0\,(t_1 s_0 F' + s_1 t_0 F) \\ H - s_0 s_1 F' + t_0 t_1 F & 0 \\ 0 & H - s_0 s_1 F' + t_0 t_1 F \end{array} \right\} \tag{2.8}$$

The next process in our calculation is to obtain the kinetic energy in a similar form and to set up the equations of motion. From these equations we derive the determinantal or secular equation, the roots of which will give the desired normal frequencies.

The kinetic energy is expressed in Cartesian coordinates $x_i$ as:

$$T = \frac{1}{2} \sum_i m_i \dot{x}_i^2 \qquad (i = 1, 2, 3, \ldots, 3N) \tag{2.9}$$

where $m_i$ is the mass of the atom associated with $x_i$ and $N$ is the number of atoms in the molecule. The internal coordinates $R_k$ of Eq. (2.7) in which we have expressed the potential energy are linear combinations of the Cartesian coordinates:

$$R_k = \sum_i B_{ki} x_i \tag{2.10}$$

where $B_{ki}$ can be obtained from geometrical considerations as follows:

$\Delta R = \Delta z_2 - \Delta z_3,$

$\Delta r = \sin \alpha_0 \cos \dfrac{\theta}{2} (\Delta x_1 - \Delta x_2) - \sin \alpha_0 \sin \dfrac{\theta}{2} (\Delta y_1 - \Delta y_2) - \cos \alpha_0 (\Delta z_1 - \Delta z_2),$

$\Delta r' = \sin \alpha_0 \cos \dfrac{\theta}{2} (\Delta x_4 - \Delta x_3) + \sin \alpha_0 \sin \dfrac{\theta}{2} (\Delta y_4 - \Delta y_3) + \cos \alpha_0 (\Delta z_4 - \Delta z_3),$

$$r_a \Delta \alpha = \rho_0 \cos \alpha_0 \cos \frac{\theta}{2} \Delta x_1 + (\rho - \rho_0 \cos \alpha_0) \cos \frac{\theta}{2} \Delta x_2 - \rho \cos \frac{\theta}{2} \Delta x_3$$

$$- \rho_0 \cos \alpha_0 \sin \frac{\theta}{2} \Delta y_1 - (\rho - \rho_0 \cos \alpha_0) \sin \frac{\theta}{2} \Delta y_2 + \rho \sin \frac{\theta}{2} \Delta y_3$$

$$+ \rho_0 \sin \alpha_0 \Delta z_1 - \rho_0 \sin \alpha_0 \Delta z_2,$$

$$r_a \Delta \alpha' = \rho_0 \cos \alpha_0 \cos \frac{\theta}{2} \Delta x_4 + (\rho - \rho_0 \cos \alpha_0) \cos \frac{\theta}{2} \Delta x_3 - \rho \cos \frac{\theta}{2} \Delta x_2$$

$$+ \rho_0 \cos \alpha_0 \sin \frac{\theta}{2} \Delta y_4 + (\rho - \rho_0 \cos \alpha_0) \sin \frac{\theta}{2} \Delta y_3 - \rho \sin \frac{\theta}{2} \Delta y_2 \quad (2.11)$$

$$- \rho_0 \sin \alpha_0 \Delta z_4 + \rho_0 \sin \alpha_0 \Delta z_3.$$

In addition to the internal motions a molecule will exert translational and rotational motions as a whole. If we express the coordinates of these motions by $X_h$, then we have

$$X_h = \sum_i B_{x_{hi}} x_i \qquad (h = 1, 2, 3, 4, 5, 6). \tag{2.12}$$

Accordingly, if we try to express $x_i$ of Eq. (2.9) in terms of $R_k$ for a molecule without translation and rotation, we meet the complicated problem of solving the simultaneous equations (2.10) and (2.12).

To overcome this difficulty Wilson[4] proposed a method which considerably reduces the labor of calculating the normal vibrations. In this method the kinetic energy is expressed in terms of momentum $p = \partial T / \partial \dot{q}$. In Cartesian coordinates we have

$$p_{x_i} = \frac{\partial T}{\partial \dot{x}_i} = m_i \dot{x}_i \tag{2.13}$$

$$T = \frac{1}{2} \sum_i \frac{1}{m_i} p_{x_i}^2 \tag{2.14}$$

Let $P_k$ and $P_{x_h}$ be, respectively, the momenta of the internal motions and of the translational and rotational motions, then we have

$$P_k = \frac{\partial T}{\partial \dot{R}_k} \tag{2.15}$$

$$P_{x_h} = \frac{\partial T}{\partial \dot{X}_h} \tag{2.16}$$

In terms of these momenta, Eq. (2.13) can be rewritten as

$$p_{x_i} = \frac{\partial T}{\partial \dot{x}_i} = \sum_k \frac{\partial T}{\partial \dot{R}_k} \frac{\partial R_k}{\partial x_i} + \sum_h \frac{\partial T}{\partial \dot{X}_h} \frac{\partial X_h}{\partial x_i} = \sum_k P_k B_{ki} + \sum_h P_{x_h} B_{x_{hi}}$$

$$\left( \begin{array}{l} k = 1, 2, \ldots\ldots, 3N - 6 \\ h = 1, 2, \ldots\ldots, 6 \end{array} \right) \tag{2.17}$$

Let us consider that the molecule is at rest so that

$$P_{x_h} = 0. \tag{2.18}$$

The second term on the right-hand side of Eq. (2.17) vanishes and the kinetic energy of Eq. (2.14) can be expressed as:

$$T = \frac{1}{2} \sum_i \frac{1}{m_i} \left( \sum_k P_k B_{ki} \right)^2 = \frac{1}{2} \sum_{k,l} \left( \sum_i \frac{1}{m_i} B_{ki} B_{li} \right) P_k P_l = \frac{1}{2} \sum_{k l} G_{kl} P_k P_l \tag{2.19}$$

where

$$G_{kl} = \sum_i B_{ki} B_{li} \frac{1}{m_i} \tag{2.20}$$

is an element of matrix $G$.

Thus we can express potential and kinetic energies in terms of internal coordinates, in which the equations of motion in Hamiltonian form are written as:

4. E. B. Wilson, J. Chem. Phys., **7**, 1047 (1939); **9**, 76 (1941).

$$\dot{R}_k = \partial(T + V)/\partial P_k \tag{2.21}$$
$$\dot{P}_k = -\partial(T + V)/\partial R_k.$$

Accordingly

$$\dot{R}_k = \sum_l G_{kl} P_l,$$

$$\dot{P}_k = -\sum_l F_{kl} R_l. \tag{2.22}$$

The vibrations being assumed to be harmonic, we have the solutions of the following form:

$$R_l = A_l \cos 2\pi\nu t, \tag{2.23}$$
$$P_l = B_l \sin 2\pi\nu t,$$

where $\nu$ is a normal frequency. Introducing these expressions into Eq. (2.22), we have

$$2\pi\nu A_k + \sum_l G_{kl} B_l = 0$$

$$\sum_l F_{kl} A_l + 2\pi\nu B_k = 0$$

$$k = 1, 2, 3, \ldots, 3N - 6.$$

For the existence of non-trivial solution of these equations it is necessary that the determinant of the coefficients vanishes:

$$\begin{vmatrix} 2\pi\nu & 0 & 0 & \cdots & G_{11} & G_{12} & G_{13} & \cdots \\ 0 & 2\pi\nu & 0 & \cdots & G_{21} & G_{22} & G_{23} & \cdots \\ 0 & 0 & 2\pi\nu & \cdots & G_{31} & G_{32} & G_{33} & \cdots \\ \cdots & \cdots & \cdots & \cdots & \cdots & \cdots & \cdots & \cdots \\ F_{11} & F_{12} & F_{13} & \cdots & 2\pi\nu & 0 & 0 & \cdots \\ F_{21} & F_{22} & F_{23} & \cdots & 0 & 2\pi\nu & 0 & \cdots \\ F_{31} & F_{32} & F_{33} & \cdots & 0 & 0 & 2\pi\nu & \cdots \\ \cdots & \cdots & \cdots & \cdots & \cdots & \cdots & \cdots & \cdots \end{vmatrix} = 0. \tag{2.24}$$

This secular equation can be rewritten as:

$$\begin{vmatrix} \sum G_{1i} F_{i1} - 4\pi^2\nu^2 & \sum G_{1i} F_{i2} & \cdots \\ \sum G_{2i} F_{i1} & \sum G_{2i} F_{i2} - 4\pi^2\nu^2 & \cdots \\ \sum G_{3i} F_{i1} & \sum G_{3i} F_{i2} & \cdots \\ \cdots & \cdots & \cdots \end{vmatrix} = 0 \tag{2.25}$$

or more simply as:

$$|\boldsymbol{G}\,\boldsymbol{F} - \boldsymbol{E}\cdot 4\pi^2\,v^2| = 0 \qquad (2.26)$$

where $\boldsymbol{E}$ is the unit matrix.

The $G$-matrix for 1,2-dihalogenoethane can be calculated from Eqs. (2.10), (2.11) and (2.20) as follows:

From Eqs. (2.8) and (2.27) we see that some of the elements of $F$-matrix and of $G$-matrix are equal and both matrices can be expressed in the following form. This is due to the presence of $C_2$ symmetry in the 1,2-dihalogenoethane molecules irrespective of the azimuthal angle of internal rotation.

$$\begin{Bmatrix} A_0 & C & C & K_0 & K_0 \\ C & A & 0 & K & T \\ C & 0 & A & T & K \\ K_0 & K & T & B & S \\ K_0 & T & K & S & B \end{Bmatrix}$$

If we replace the internal coordinates, $\Delta R,\ \Delta r,\ \Delta r'$, $r_a\Delta\alpha$ and $r_a\Delta\alpha'$ by the internal symmetry coordinates:*

$$\Delta R$$
$$(\Delta r + \Delta r')/\sqrt{2}$$
$$r_a(\Delta\alpha + \Delta\alpha')/\sqrt{2}$$
$$(\Delta r - \Delta r')/\sqrt{2}$$
$$r_a(\Delta\alpha - \Delta\alpha')/\sqrt{2}$$

both of the $G$- and the $F$-matrices can be factored into two matrices of lower degrees:

$$\begin{Bmatrix} A_0 & \sqrt{2}\,C & \sqrt{2}\,K_0 \\ \sqrt{2}\,C & A & K+T \\ \sqrt{2}\,K_0 & K+T & B+S \end{Bmatrix}$$

and

$$\begin{Bmatrix} A & K-T \\ K-T & B-S \end{Bmatrix}$$

---

* The internal symmetry coordinates are linear combinations of internal coordinates and correspond to displacements of the atoms in agreement with one of the symmetry types.

$$G = \begin{pmatrix} 2\mu_0 & \mu_0\cos\alpha_0 & \mu_0\cos\alpha_0 & -\rho_0\mu_0\sin\alpha_0 & -\rho_0\mu_0\sin\alpha_0 \\ \mu_0\cos\alpha_0 & \mu+\mu_0 & 0 & -\rho\mu_0\sin\alpha_0 & \rho\mu_0\sin\alpha_0\cos\theta \\ \mu_0\cos\alpha_0 & 0 & \mu+\mu_0 & \rho\mu_0\sin\alpha_0\cos\theta & -\rho\mu_0\sin\alpha_0 \\ -\rho_0\mu_0\sin\alpha_0 & -\rho\mu_0\sin\alpha_0 & \rho\mu_0\sin\alpha_0\cos\theta & \rho_0^2\mu+(2\rho^2+\rho_0^2-2\cos\alpha_0)\mu_0 & -2\mu_0(\rho^2-\cos\alpha_0)\cos\theta \\ -\rho_0\mu_0\sin\alpha_0 & \rho\mu_0\sin\alpha_0\cos\theta & -\rho\mu_0\sin\alpha_0 & -2\mu_0(\rho^2-\cos\alpha_0)\cos\theta & \rho_0^2\mu+(2\rho^2+\rho_0^2-2\cos\alpha_0)\mu_0 \end{pmatrix} \qquad (2.27)$$

Accordingly we have from Eqs. (2.8) and (2.27) the following relations:

$$G_A =$$
$$\begin{pmatrix} 2\mu_0 & \sqrt{2}\,\mu_0 \cos\alpha_0 & -\sqrt{2}\,\rho_0\mu_0 \sin\alpha_0 \\ \sqrt{2}\,\mu_0 \cos\alpha_0 & \mu+\mu_0 & -\rho\mu_0 \sin\alpha_0 \,(1-\cos\theta) \\ -\sqrt{2}\,\rho_0\mu_0 \sin\alpha_0 & -\rho\mu_0 \sin\alpha_0\,(1-\cos\theta) & \rho_0{}^2\,(\mu+\mu_0)+2\mu_0\,(\rho^2-\cos\alpha_0)\,(1-\cos\theta) \end{pmatrix}$$
$$(2.28)$$

$$F_A = \begin{pmatrix} k_R + 2t_0{}^2 F' + 2s_0{}^2 F & \sqrt{2}\,(-t_0\,t_1\,F'+s_0\,s_1\,F) & \sqrt{2}\,\rho\,(t_0\,s_1\,F'+s_0\,t_1\,F) \\ \sqrt{2}\,(-t_0\,t_1\,F'+s_0\,s_1\,F) & K_r + t_1{}^2 F' + s_1{}^2 F & \rho_0\,(t_1\,s_0\,F'+s_1\,t_0\,F) \\ \sqrt{2}\,\rho\,(t_0\,s_1\,F'+s_0\,t_1\,F) & \rho_0\,(t_1\,s_0\,F'+s_1\,t_0\,F) & H - s_0\,s_1\,F' + t_0\,t_1\,F \end{pmatrix}$$
$$(2.29)$$

$$G_B = \begin{pmatrix} \mu+\mu_0 & -\rho\mu_0 \sin\alpha_0\,(1+\cos\theta) \\ -\rho\mu_0 \sin\alpha_0\,(1+\cos\theta) & \rho_0{}^2\,(\mu+\mu_0)+2\mu_0\,(\rho^2-\cos\alpha_0)\,(1+\cos\theta) \end{pmatrix}$$
$$(2.30)$$

$$F_B = \begin{pmatrix} K_r + t_1{}^2 F' + s_1{}^2 F & \rho_0\,(t_1\,s_0\,F'+s_1\,t_0\,F) \\ \rho_0\,(t_1\,s_0\,F'+s_1\,t_0\,F) & H - s_0\,s_1\,F' + t_0\,t_1\,F \end{pmatrix} \qquad (2.31)$$

By solving the secular equation containing $G_A$ and $F_A$ (corresponding to the symmetric coordinates) the C–C stretching frequency $\nu$ (C–C), the symmetric C–X stretching frequency $\nu_s$ (C–X) and the symmetric C–C–X deformation frequency $\delta_s$ (C–C–X) are obtained. From the other secular equation containing $G_B$ and $F_B$ (corresponding to the antisymmetric

TABLE 2.1

Force constants, bond lengths and bond angle of the skeleton of 1,2-dihalogenoethanes

|  | Dichloroethane | Dibromoethane |
|---|---|---|
| $r_0$ | 1.76Å | 1.92Å |
| $R_0$ | 1.54Å | 1.54Å |
| $\alpha_0$ | tetrahedral angle | tetrahedral angle |
| $K_R$ | $3.7 \times 10^5$ dynes/cm. | $3.7 \times 10^5$ dynes/cm. |
| $K_r$ | $2.9 \times 10^5$ dynes/cm. | $2.4 \times 10^5$ dynes/cm. |
| $H$ | $0.17 \times 10^5$ dynes/cm. | $0.15 \times 10^5$ dynes/cm. |
| $F$ | $0.49 \times 10^5$ dynes/cm. | $0.40 \times 10^5$ dynes/cm. |
| $F'$ | $-0.1F \times 10^5$ dynes/cm. | $-0.1F \times 10^5$ dynes/cm. |

coordinates) the antisymmetric C–X stretching frequency $v_a$ (C–X) and the antisymmetric C–C–X deformation frequency $\delta_a$ (C–C–X) are obtained. With the values of atomic distances, bond angles and the force constants shown in Table 2.1, we can calculate these frequencies for the *trans* and *gauche* forms as shown in Table 2.2. The agreement between the calculated and observed frequencies is good, if we consider that the calculation deals with the pure skeletal vibrations and therefore, it does not take into account the coupling with the hydrogen vibrations or with the torsional oscillation about the C–C axis.

TABLE 2.2

Calculated and observed skeletal frequencies of 1,2-dihalogenoethanes (cm.$^{-1}$)

| Configuration | Type of vibration | Dichloroethane | | Dibromoethane | |
|---|---|---|---|---|---|
| | | Calc. | Obs. | Calc. | Obs. |
| *trans* | $v_s$ (C–C) | 1008 | 1052 | 994 | 1053 |
| | $v_s$ (C–X) | 745 | 754 | 662 | 660 |
| | $\delta_s$ (C–C–X) | 287 | 300 | 182 | 190 |
| | $v_a$ (C–X) | 741 | 709 | 620 | 587 |
| | $\delta_a$ (C–C–X) | 226 | 223 | 181 | — |
| *gauche* | $v_1$ (C–C) | 1006 | 1031 | 994 | 1019 |
| | $v_1$ (C–X) | 674 | 654 | 564 | 551 |
| | $\delta_1$ (C–C–X) | 252 | 265 | 209 | 231 |
| | $v_2$ (C–X) | 738 | 677 | 634 | 583 |
| | $\delta_2$ (C–C–X) | 421 | 411 | 353 | 355 |

## 7. Normal Vibrations of *n*-Paraffins

In Section 19 of Part I we have calculated the skeletal frequencies of *n*-paraffin molecules in the extended form, which is the only stable form in the solid state. There we were interested in discussing a deformation frequency which changes in a characteristic manner with the length of chain (from 425 cm.$^{-1}$ for butane to 150 cm.$^{-1}$ for cetane). This frequency is one of the two frequencies corresponding to the condition

$$\lambda' = \pi/N$$

of Eq. (5.12), Part I where $N$ is the number of carbon atoms.  The other one corresponds to a Raman line which appears in the frequency region $800 \sim 900$ cm.$^{-1}$ (837 cm.$^{-1}$ for butane to 888 cm.$^{-1}$ for cetane).

As shown in Table 5.2 of Part I there are observed other Raman lines whose frequencies remain almost constant throughout the homologous series.  In the case of solid cetane $C_{16}H_{34}$ these frequencies are 1058, 1135, 1295, 1442, 1471, 2846, 2878, 2934 and 2963 cm.$^{-1}$.  We shall discuss, in the following, the nature of these frequencies, taking into account the motions of hydrogen atoms  (Shimanouchi and Mizushima)[5].

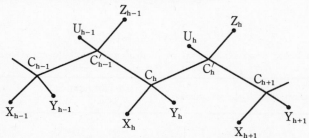

Fig. 2.3.  A part of the extended chain of $-(CXY-CZU)_n-$.

A part of an infinite extended chain,

$$- (C\,X\,Y - C'\,Z\,U)_n -$$

is shown in Fig. 2.3, from which the meaning of the following  internal coordinates will be clear.  $R_0, r_{10}, r_{20}, \ldots$ indicated in parentheses denote the equilibrium values.

$$
\begin{aligned}
&R_h\,(C_h - C_h{'} = R_0), &\qquad &R_h{'}\,(C_h{'} - C_{h+1} = R_0{'}) \\
&r_{1h}\,(C_h - X_h = r_{10}) & &r_{1h}{'}\,(C_h{'} - Z_h = r_{10}{'}), \\
&r_{2h}\,(C_h - Y_h = r_{20}), & &r_{2h}{'}\,(C_h{'} - U_h = r_{20}{'}), \\
&\Theta_h\,(\measuredangle\,C_{h-1}{'}\,C_h\,C_h{'} = \beta), & &\Theta_h{'}\,(\measuredangle\,C_h\,C_h{'}\,C_{h+1} = \beta{'}), \\
&\theta_h\,(\measuredangle\,X_h\,C_h\,Y_h = \alpha), & &\theta_h{'}\,(\measuredangle\,Z_h\,C_h{'}\,U_h = \alpha{'}), \\
&\varphi_{1h}\,(\measuredangle\,X_h\,C_h\,C_h{'} = \gamma_1), & &\varphi_{1h}{'}\,(\measuredangle\,Z_h\,C_h{'}\,C_h = \gamma_1{'}), \\
&\varphi_{2h}\,(\measuredangle\,Y_h\,C_h\,C_h{'} = \gamma_2), & &\varphi_{2h}{'}\,(\measuredangle\,U_h\,C_h{'}\,C_h = \gamma_2{'}), \\
&\varphi_{3h}\,(\measuredangle\,X_h\,C_h\,C_{h-1}{'} = \gamma_3), & &\varphi_{3h}{'}\,(\measuredangle\,Z_h\,C_h{'}\,C_{h+1} = \gamma_3{'}), \\
&\varphi_{4h}\,(\measuredangle\,Y_h\,C_h\,C_{h-1}{'} = \gamma_4), & &\varphi_{4h}{'}\,(\measuredangle\,U_h\,C_h{'}\,C_{h+1} = \gamma_4{'}).
\end{aligned}
$$

The suffix $h$ of these 18 kinds of coordinates changes from zero to infinity for an infinite chain.

By use of these internal coordinates the $G$-matrices related to the kinetic energy can be constructed.  There are in all $18 \times 18 = 324$ such matrices of $\infty$ th order.  We shall only show a part of them.

5.  T. Shimanouchi and S. Mizushima, J. Chem. Phys., 17, 1102 (1949).

$$
\begin{array}{c|ccccccccccccccc}
 & \cdots & \Delta R_{h-1} & \Delta R_{h} & \Delta R_{h+1} & \cdots & \cdots & \Delta R_{h-1}' & \Delta R_{h}' & \Delta R_{h+1}' & \cdots & \cdots & \Delta r_{1,h-1} & \Delta r_{1,h} & \Delta r_{1,h+1} & \cdots \\
\hline
\vdots & & \vdots & \vdots & \vdots & & & \vdots & \vdots & \vdots & & & \vdots & \vdots & \vdots & \\
\Delta R_{h-1} & & A_0 & 0 & 0 & & & C_0' & 0 & 0 & & & C_{01} & 0 & 0 & \\
\Delta R_{h} & & 0 & A_0 & 0 & & & C_0 & C_0' & 0 & & & 0 & C_{01} & 0 & \\
\Delta R_{h+1} & & 0 & 0 & A_0 & & & 0 & C_0 & C_0' & & & 0 & 0 & C_{01} & \\
\vdots & & \vdots & \vdots & \vdots & & & \vdots & \vdots & \vdots & & & \vdots & \vdots & \vdots & \\
\end{array}
\qquad (2.32)
$$

Here $A_0$, $C_0$, $C_0'$, $C_{01}$, etc., are determined from the molecular configuration just as in the case of the 1,2-dihalogenoethanes explained in the preceding section:

$$A_0 = \mu_0 + \mu_0',$$
$$C_0 = \mu_0 \cos \beta,$$
$$C_0' = \mu_0' \cos \beta',$$
$$C_{01} = \mu_0 \cos \gamma_1,$$
$$\ldots\ldots\ldots\ldots,$$

where $\mu_0$ and $\mu_0'$ denote the reciprocals of masses of $C_h$- and $C_h'$-atoms, respectively.

Assuming the Urey-Bradley field and neglecting the interaction between atoms far apart, we express the potential energy as:

$$
\begin{aligned}
V_h = \sum_{R,\,R'} & \left[ K_R' R_0 (\Delta R_h) + \frac{1}{2} K_R (\Delta R_h)^2 \right] \\
+ \sum_{r_1, r_2, r_1', r_2'} & \left[ K_r' r_{10} (\Delta r_{1h}) + \frac{1}{2} K_r (\Delta r_{1h})^2 \right] \\
+ \sum_{\Theta,\,\Theta'} & \left[ H_\Theta' R_0^2 (\Delta \Theta_h) + \frac{1}{2} H_\Theta (R_0 \Delta \Theta_h)^2 \right] \\
+ \sum_{\theta,\,\theta'} & \left[ H_\theta' r_{10} r_{20} (\Delta \theta_h) + \frac{1}{2} H_\theta r_{10} r_{20} (\Delta \theta_h)^2 \right] \\
+ \sum_{\varphi_1, \varphi_2, \varphi_3, \varphi_4, \varphi_1', \varphi_2', \varphi_3', \varphi_4'} & \left[ H_\varphi' r_{10} R_0 (\Delta \varphi_{1h}) \right. \\
& \left. + \frac{1}{2} H_\varphi r_{10} R_0 (\Delta \varphi_{1h})^2 \right] + \sum_{Q,\,Q'} \left[ F_Q' Q_0 (\Delta Q_h) \right. \\
& \left. + \frac{1}{2} F_Q (\Delta Q_h)^2 \right] + \sum_{q,\,q'} \left[ F_q' q_0 (\Delta q_h) + \frac{1}{2} F_q (\Delta q_h)^2 \right] \\
+ \sum_{p_1, p_2, p_3, p_4, p_1', p_2', p_3', p_4'} & \left[ F_p' p_{10} (\Delta p_{1h}) + \frac{1}{2} F_p (\Delta p_{1h})^2 \right]
\end{aligned}
\tag{2.33}
$$

where $Q_h$, $Q_h'$, $q_h$, $q_h'$, $p_{1h} \cdots p_{4h}'$ denote the following interatomic distances not bonded directly with equilibrium values indicated in parentheses.

$$
\begin{array}{ll}
Q_h \, (C_{h-1}' \ldots C_h' = Q_0), & Q_h' \, (C_h \ldots C_{h+1} = Q_0'), \\
q_h \, (X_h \ldots Y_h = q_0), & q_h' \, (Z_h \ldots U_h = q_0'), \\
p_{1h} \, (X_h \ldots C_h' = p_{10}), & p_{1h}' \, (Z_h \ldots C_h = p_{10}'), \\
p_{2h} \, (Y_h \ldots C_h' = p_{20}), & p_{2h}' \, (U_h \ldots C_h = p_{20}'), \\
p_{3h} \, (X_h \ldots C_{h-1}' = p_{30}), & p_{3h}' \, (Z_h \ldots C_{h+1} = p_{30}'), \\
p_{4h} \, (Y_h \ldots C_{h-1}' = p_{40}), & p_{4h}' \, (U_h \ldots C_{h+1} = p_{40}').
\end{array}
$$

These can be replaced by the previously-described internal coordinates $\Delta R_h, \Delta R_h', \Delta r_{1h}, \Delta r_{1h}' \ldots \Delta \varphi_{4h}'$ and, therefore, the potential energy $V$ can also be expressed in a quadratic form in terms of these internal coordinates. From this potential energy expression the $F$-matrices can at once be constructed, just as in the preceding section.

For the selection rule for a long chain molecule we have to consider the translational symmetry as well as the rotational symmetry. For the former, so long as the chain is short, a vibration will be active for

$$\lambda' = \pi/N \tag{2.34}$$

but for a long chain we must have

$$\lambda' = 0 \tag{2.35}$$

in other words all the atoms vibrate in phase (totally symmetric vibration) Corresponding to this vibration we have a secular equation of 18-th order

$$|\boldsymbol{G}\,\boldsymbol{F} - \boldsymbol{E}\,\lambda| = 0, \tag{2.36}$$

which is obtained by the reduction of the $G$- and $F$-matrices of $\infty$ th order stated above. The elements of matrices of this secular equation can readily be obtained by summing all the elements in a row or a column of the matrices of $\infty$ th order. For example from the $G$-matrix shown above we have a new matrix:

|  | $\Delta R$ | $\Delta R'$ | $\Delta r_1$ | $\cdots$ | $\cdots$ |
|---|---|---|---|---|---|
| $\Delta R$ | $A_0$ | $C_0 + C_0'$ | $C_{01}$ | $\cdots$ | $\cdots$ |
| $\cdots$ | $\cdots$ | $\cdots$ | $\cdots$ | $\cdots$ | $\cdots$ |

For the molecule of $n$-paraffin we can put the equilibrium values of interatomic distances and bond angles as:

$$R_0 = R_0' = 1.54\text{Å}, \qquad r_{10} = r_{10}' = r_{20} = r_{20}' = r_0 = 1.09\text{Å},$$

$$\alpha = \alpha' = \beta = \beta' = \gamma_1 = \gamma_1' = \gamma_2 = \gamma_2' = \gamma_3 = \gamma_3' = \gamma_4 = \gamma_4' = \text{tetrahedral angle}$$

We have then the $G$- and $F$-matrices of 18th order of the form as shown in Table 2.3.

If we express the secular equation as

$$|\boldsymbol{G}\,\boldsymbol{F} - 3\,M_c\,\boldsymbol{E}\,\lambda| = 0, \tag{2.37}$$

(where $M_c$ denotes the mass of a carbon atom), the elements of the $G$-matrix become:

$a_0 = 6,$

$a = 3 + w,$

$b_0 = 32\,v^2,$

$b = 16 + 4\,w,$

$b_1 = 12\,v^2 + 4\,v + 6 + 2\,w,$

$c_0 = -2,$

$c = -1,$

$c_1 = -1,$

$d_0 = -16\,v,$

$d_1 = -4\,(2\,v^2 + 2\,v + 1),$

$e_0 = -8\,v^2 + 4\,v,$

$e = 4\,v - 4 - w,$

$e_1 = -6\,v^2 - 2\,v + 5$

$e_2 = 10\,v^2 - 2\,v - 3 - w,$

$g_0 = -8\,v^2 + 4\,v,$

$g = 32\,v^2,$

$g_1 = 12\,v^2 + 4\,v,$

$g_2 = -6\,v^2 - 2\,v,$

$h = 4\,v,$

$h_1 = -4\,v,$

$h_2 = 2\,v,$

$k_0 = -8\,v,$

$k = -4,$

$k_1 = -4,$

$k_2 = -4\,v,$

$l_0 = 4,$

$l = 4\,v,$

$l_1 = 4\,v + 2,$

$l_2 = 2\,v + 2,$

$u_0 = -16\,v,$

$u = 4\,v,$

$u_1 = 10\,v^2 - 2\,v,$

$u_2 = -8\,v^2 - 8\,v,$

where $v = r_0/R_0$ and $w = 3\,M_c/M_H$, $M_H$ denoting the mass of a hydrogen atom.

TABLE 2.3

The $F$- and $G$-matrices for the infinite chain of $n$-paraffin

| | | | | | | | | | | | | | | | | | | | |
|---|---|---|---|---|---|---|---|---|---|---|---|---|---|---|---|---|---|---|---|
| $\Delta R$ | $a_0$ | | | | | | | | | | | | | | | | | | |
| $\Delta R'$ | $c_0$ | $a_0$ | | | | | | | | | | | | | | | | | |
| $\Delta r_1$ | $c_1$ | $c_1$ | $a$ | | | | | | | | | | | | | | | | |
| $\Delta r_2$ | $c_1$ | $c_1$ | $c$ | $a$ | | | | | | | | | | | | | | | |
| $\Delta r_1'$ | $c_1$ | $c_1$ | $0$ | $0$ | $a$ | | | | | | Symmetrical matrix | | | | | | | | |
| $\Delta r_2'$ | $c_1$ | $c_1$ | $0$ | $0$ | $c$ | $a$ | | | | | | | | | | | | | |
| $\sqrt{2}\Delta\Theta$ | $k_0$ | $k_0$ | $l$ | $l$ | $h$ | $h$ | $b_0$ | | | | | | | | | | | | |
| $\sqrt{2}\Delta\Theta'$ | $k_0$ | $k_0$ | $h$ | $h$ | $l$ | $l$ | $g$ | $b_0$ | | | | | | | | | | | |
| $\sqrt{2}\Delta\theta$ | $l_0$ | $l_0$ | $k$ | $k$ | $0$ | $0$ | $d_0$ | $u_0$ | $b$ | | | | | | | | | | |
| $\sqrt{2}\Delta\theta'$ | $l_0$ | $l_0$ | $0$ | $0$ | $k$ | $k$ | $u_0$ | $d_0$ | $0$ | $b$ | | | | | | | | | |
| $\sqrt{2}\Delta\varphi_1$ | $k_1$ | $l_1$ | $k_2$ | $l_2$ | $h_1$ | $h_2$ | $e_0$ | $g_0$ | $e$ | $u$ | $b_1$ | | | | | | | | |
| $\sqrt{2}\Delta\varphi_2$ | $k_1$ | $l_1$ | $l_2$ | $k_2$ | $h_2$ | $h_1$ | $e_0$ | $g_0$ | $e$ | $u$ | $e_1$ | $b_1$ | | | | | | | |
| $\sqrt{2}\Delta\varphi_3$ | $l_1$ | $k_1$ | $k_2$ | $l_2$ | $h_1$ | $h_2$ | $e_0$ | $g_0$ | $e$ | $u$ | $e_2$ | $d_1$ | $b_1$ | | | | | | |
| $\sqrt{2}\Delta\varphi_4$ | $l_1$ | $k_1$ | $l_2$ | $k_2$ | $h_2$ | $h_1$ | $e_0$ | $g_0$ | $e$ | $u$ | $d_1$ | $e_2$ | $e_1$ | $b_1$ | | | | | |
| $\sqrt{2}\Delta\varphi_1'$ | $k_1$ | $l_1$ | $h_1$ | $h_2$ | $k_2$ | $l_2$ | $g_0$ | $e_0$ | $u$ | $e$ | $g_1$ | $g_2$ | $u_1$ | $u_2$ | $b_1$ | | | | |
| $\sqrt{2}\Delta\varphi_2'$ | $k_1$ | $l_1$ | $h_2$ | $h_1$ | $l_2$ | $k_2$ | $g_0$ | $e_0$ | $u$ | $e$ | $g_2$ | $g_1$ | $u_2$ | $u_1$ | $e_1$ | $b_1$ | | | |
| $\sqrt{2}\Delta\varphi_3'$ | $l_1$ | $k_1$ | $h_1$ | $h_2$ | $k_2$ | $l_2$ | $g_0$ | $e_0$ | $u$ | $e$ | $u_1$ | $u_2$ | $g_1$ | $g_2$ | $e_2$ | $d_1$ | $b_1$ | | |
| $\sqrt{2}\Delta\varphi_4'$ | $l_1$ | $k_1$ | $h_2$ | $h_1$ | $l_2$ | $k_2$ | $g_0$ | $e_0$ | $u$ | $e$ | $u_2$ | $u_1$ | $g_2$ | $g_1$ | $d_1$ | $e_2$ | $e_1$ | $b_1$ | |

For the elements of the $F$-matrix we have:

$$a_0 = K_{CC} + \frac{2}{3}F_{cc}' + \frac{4}{3}F_{CC} + 8t_0{}^2 F_{CH}' + 4s_0{}^2 F_{CH},$$

$$a = K_{CH} + \frac{1}{3}F_{HH}' + \frac{2}{3}F_{HH} + 4t_1{}^2 F_{CH}' + 2s_1{}^2 F_{CH},$$

$$c_0 = 2\left(-\frac{1}{3}F_{CC}' + \frac{2}{3}F_{CC}\right),$$

$$c = -\frac{1}{3}F_{HH}' + \frac{2}{3}F_{HH},$$

$$c_1 = -2t_0 t_1 F_{CH}' + s_0 s_1 F_{CH},$$

$$k_0 = \left(\frac{1}{3}F_{CC}' + \frac{1}{3}F_{CC}\right)\Big/ v,$$

$$k = \frac{1}{3}(F_{HH}' + F_{HH}),$$

$$k_1 = t_0 s_1 F_{CH}' + t_1 s_0 F_{CH},$$

$$k_2 = (t_1 s_0 F_{CH}' + t_0 s_1 F_{CH})/v,$$

$$2b_0 = \left(H_{CCC} - \frac{2}{3}F_{CC}' + \frac{1}{3}F_{CC}\right)\Big/ v^2 + (3\varkappa/2\sqrt{2}\,r_0{}^2),$$

$$2b = \left(H_{HCH} - \frac{2}{3}F_{HH}' + \frac{1}{3}F_{HH}\right) + (3\varkappa/2\sqrt{2}\,r_0{}^2),$$

$$2b_1 = (H_{CCH} - s_0 s_1 F_{CH}' + 2t_0 t_1 F_{CH})/v + (3\varkappa/2\sqrt{2}\,r_0{}^2),$$

$$e_0 = e = e_1 = e_2 = \varkappa/2\sqrt{2}\,r_0{}^2,$$

$$l_0 = l = l_1 = l_2 = d_0 = d_1 = g_0 = g_1 = g_2 = g = h = h_1 = h_2 = u_0 = u = u_1 = u_2 = 0.$$

where $\varkappa$ denotes the intramolecular tension[6] and

$$s_0 = \left(R_0 + \frac{1}{3}r_0\right)\Big/ p_{10}, \qquad s_1 = \left(r_0 + \frac{1}{3}R_0\right)\Big/ p_{10}, \qquad t_0 = \frac{2}{3}r_0/p_{10},$$

$$t_1 = \frac{2}{3}R_0/p_{10}$$

Taking into account the symmetry of the matrix shown in Table 2.3, we shall use the following symmetry coordinates which are linear combinations of the internal coordinates.*

* The symmetry coordinates of the $A_{1g}$ and $A_{1u}$ types contain one redundant coordinate each.

6. T. Shimanouchi, J. Chem. Phys., **17**, 245 (1949).

$A_{1g}$:
$$(\Delta R + \Delta R')/\sqrt{2},$$
$$(\Delta r_1 + \Delta r_2 + \Delta r_1' + \Delta r_2')/2,$$
$$(\Delta \Theta + \Delta \Theta')/\sqrt{2},$$
$$(\Delta \theta + \Delta \theta')/\sqrt{2},$$
$$(\Delta \varphi_1 + \Delta \varphi_2 + \Delta \varphi_3 + \Delta \varphi_4 + \Delta \varphi_1' + \Delta \varphi_2' + \Delta \varphi_3' + \Delta \varphi_4')/(8)^{1/2}.$$

$A_{2g}$: $\quad (\Delta \varphi_1 - \Delta \varphi_2 - \Delta \varphi_3 + \Delta \varphi_4 + \Delta \varphi_1' - \Delta \varphi_2' - \Delta \varphi_3' + \Delta \varphi_4')/(8)^{1/2},$

$B_{1g}$:
$$(\Delta R - \Delta R')/\sqrt{2},$$
$$(\Delta \varphi_1 + \Delta \varphi_2 - \Delta \varphi_3 - \Delta \varphi_4 + \Delta \varphi_1' + \Delta \varphi_2' - \Delta \varphi_3' - \Delta \varphi_4')/(8)^{1/2}$$

$B_{2g}$:
$$(\Delta r_1 - \Delta r_2 + \Delta r_1' - \Delta r_2')/2,$$
$$(\Delta \varphi_1 - \Delta \varphi_2 + \Delta \varphi_3 - \Delta \varphi_4 + \Delta \varphi_1' - \Delta \varphi_2' + \Delta \varphi_3' - \Delta \varphi_4')/(8)^{1/2}$$

$A_{1u}$:
$$(\Delta r_1 + \Delta r_2 - \Delta r_1' - \Delta r_2')/2,$$
$$(\Delta \Theta - \Delta \Theta')/\sqrt{2},$$
$$(\Delta \theta - \Delta \theta')/\sqrt{2},$$
$$(\Delta \varphi_1 + \Delta \varphi_2 + \Delta \varphi_3 + \Delta \varphi_4 - \Delta \varphi_1' - \Delta \varphi_2' - \Delta \varphi_3' - \Delta \varphi_4')/(8)^{1/2},$$

$A_{2u}$: $\quad (\Delta \varphi_1 - \Delta \varphi_2 - \Delta \varphi_3 + \Delta \varphi_4 - \Delta \varphi_1' + \Delta \varphi_2' + \Delta \varphi_3' - \Delta \varphi_4')/(8)^{1/2},$

$B_{1u}$: $\quad (\Delta \varphi_1 + \Delta \varphi_2 - \Delta \varphi_3 - \Delta \varphi_4 - \Delta \varphi_1' - \Delta \varphi_2' + \Delta \varphi_3' + \Delta \varphi_4')/(8)^{1/2},$

$B_{2u}$:
$$(\Delta r_1 - \Delta r_2 - \Delta r_1' + \Delta r_2')/2,$$
$$(\Delta \varphi_1 - \Delta \varphi_2 + \Delta \varphi_3 - \Delta \varphi_4 - \Delta \varphi_1' + \Delta \varphi_2' - \Delta \varphi_3' + \Delta \varphi_4').$$

The $G$- and $F$-matrices are then reduced, the redundant coordinates disappear, and we have finally:*

$$G_{A_{1g}} = \begin{pmatrix} 2+w, & -4, & 4 \\ -4, & 8, & -8 \\ 4, & -8, & 8+2w \end{pmatrix}$$

$$F_{A_{1g}} = \begin{pmatrix} a+c, & 2c_1 + 2v k_2, & -2k + k_2 \\ 2c_1 + 2v k_2, & \{(a_0 + c_0) - 8v(2k_0 - k_1) + 4v^2(4b_0 + b_1 - 6e)\}/2, & k_1 + v(b_1 - 6e) \\ -2k + k_2, & k_1 + v(b_1 - 6e), & (4b + b_1 - 6e)/2 \end{pmatrix}$$

---

* The $G_{A_{1g}}$ -matrix expressed in terms of these symmetry coordinates is of the fourth order. However, through a suitable transformation we have the matrix of the third order shown here. The remaining one root of the secular equation is zero. Similarly we have the $G_{A_{1u}}$ -matrix of the second order which was originally of the third order.

$$G_{A_{2g}} = (3\,w)\,,$$

$$F_{A_{2g}} = (b_1 - 2\,e)\,,$$

$$F_{A_{1u}} = \begin{pmatrix} a + c, & 2\,k - k_2, \\ 2\,k - k_2, & (4\,b + b_1 - 6\,e)/2 \end{pmatrix}$$

$$G_{B_{1g}} = \begin{pmatrix} 16, & -8\,(2\,v + 3) \\ -8\,(2\,v + 3), & 4\,(2\,v+3)^2+6\,w \end{pmatrix} \quad G_{A_{2u}} = (3\,w),$$

$$F_{B_{1g}} = \begin{pmatrix} (a_0 - c_0)/2, & k_1, \\ k_1, & b_1/2, \end{pmatrix} \qquad F_{A_{2u}} = (b_1 - 2\,e),$$

$$G_{B_{2g}} = \begin{pmatrix} 4 + w, & -4\,(6\,v + 1), \\ -4\,(6\,v + 1), & 4\,(6\,v + 1)^2 + 2\,w \end{pmatrix} \quad \begin{aligned} G_{B_{1u}} &= (18 + 3\,w), \\ F_{B_{1u}} &= (b_1), \end{aligned}$$

$$F_{B_{2g}} = \begin{pmatrix} a - c, & k_2, \\ k_2 & b_1/2 \end{pmatrix} \qquad G_{B_{2u}} = \begin{pmatrix} 4 + w, & -4, \\ -4, & 4 + 2\,w, \end{pmatrix}$$

$$G_{A_{1u}} = \begin{pmatrix} 2 + w, & -4, \\ -4 & 8 + 2\,w \end{pmatrix} \qquad F_{B_{2u}} = \begin{pmatrix} a - c, & k_2, \\ k_2 & b_1/2 \end{pmatrix}$$

As the chain is considered to be infinitely long, the extended zigzag chain has the symmetry $C_{2v}$ or $C_{2h}$. In the former case the rotation axis $C_2$ passes through a carbon nucleus and in the latter case it passes through the middle point of a bond connecting two consecutive carbon nuclei. The selection rule is shown in Table 2.4, in which $C_2$ refers to $C_{2v}$, $\sigma$ to $C_{2v}$ and $C_{2h}$ and $i$ to $C_{2h}$. We have, therefore, five infrared active frequencies and eight Raman active frequencies.[7] The nature of the displacements of the atomic nuclei in these vibrations is shown in Fig. 2.4.

For the numerical calculation of frequencies we shall use the values of force constants (shown in Table 2.5) with which we have succeeded in calculating the frequencies of molecules with similar structures. (It is to be noted that these values of force constants are different from those used in the calculation of skeletal frequencies described in Section 19, Part I. For example, the value of $K_{CC}$ of Table 2.5 is $2.8 \times 10^5$ dynes/cm., while the corresponding quantity in the calculation of the skeletal motion has a value of $4.0 \times 10^5$ dynes/cm. This is due to the fact that in the latter case we have to include in this force constant the interaction between the non-bonded C and H atoms, which corresponds to $F_{CH}$ of the present calculation.) In Table 2.6 are shown the calculated and observed frequencies, the latter being taken from the Raman spectrum of solid N-cetane (Table 5.2, Part I)

7. T. Shimanouchi, J. Chem. Phys., **17**, 734 (1949).

TABLE 2.4

Selection rule for each type of vibration

| | $C_2$ | $\sigma$ | $i$ | |
|---|---|---|---|---|
| $A_{1g}$ | + | + | + | Raman active |
| $A_{2g}$ | + | — | + | Raman active |
| $B_{1g}$ | — | + | + | Raman active |
| $B_{2g}$ | — | — | + | Raman active |
| $A_{1u}$ | + | + | — | Infrared active |
| $A_{2u}$ | + | — | — | — |
| $B_{1u}$ | — | + | — | Infrared active |
| $B_{2u}$ | — | — | — | Infrared active |

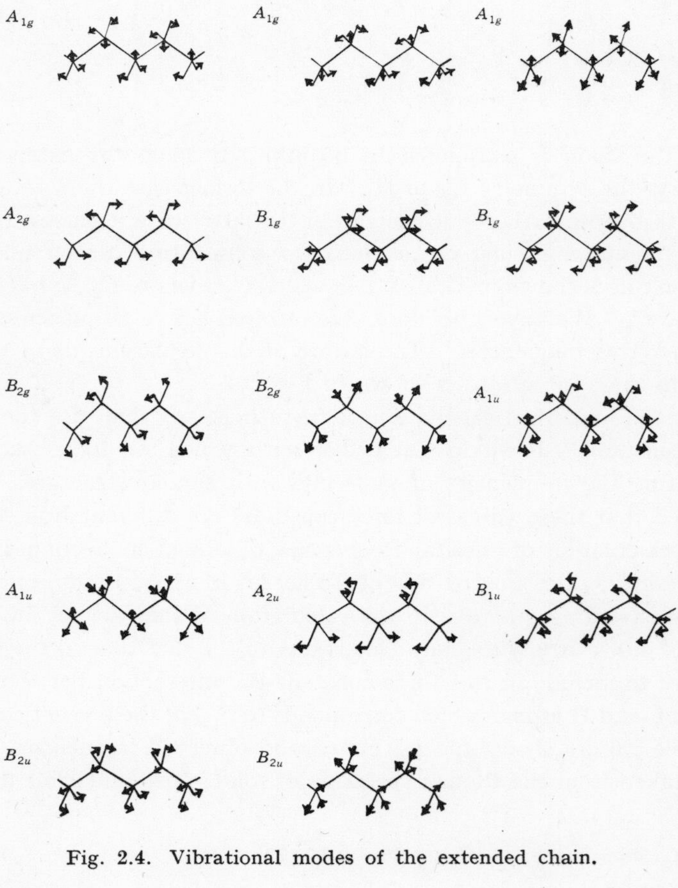

Fig. 2.4. Vibrational modes of the extended chain.

and the infrared absorption spectrum of polyethylene.  The agreement between the theoretical and experimental values is good except for the line at 1295 cm.$^{-1}$.  This line corresponds to that at 1155 cm.$^{-1}$ of ethane which is not in good agreement with the theoretical value ($E_g$) obtainable from a similar secular equation.[7]  The discrepancies in these two cases may be due to a common origin.

TABLE 2.5

Force constants in $10^5$ dynes/cm.

| $K_{CC} = 2.8$ | $H_{CCC} = 0.2$ | $F_{CC'} = -0.05$ | $F_{CC} = 0.3$ |
| $K_{CH} = 4.2$ | $H_{HCH} = 0.4$ | $F_{HH'} = 0$ | $F_{HH} = 0.1$ |
| | $H_{CCH} = 0.15$ | $F_{CH'} = -0.05$ | $F_{CH} = 0.4$ |

$(\varkappa = 0.2 \times 10^{-11}$ dynes·cm.)

TABLE 2.6

The calculated and the observed frequencies (cm.$^{-1}$).

| | Calculated | Observed | |
|---|---|---|---|
| $A_{1g}$ | 1073<br>1516<br>2873 | 1135<br>1471<br>$\{$ 2846<br>2878 | |
| $A_{2g}$ | 978 | — | $\left(\begin{array}{c}\text{Raman lines of solid}\\\text{N-cetane}\end{array}\right)$ |
| $B_{1g}$ | 1016<br>1435 | 1058<br>1442 | |
| $B_{2g}$ | 1112<br>2910 | 1295?<br>$\{$ 2934<br>2963 | |
| $A_{1u}$ | 1475<br>2907 | 1475<br>2853 | |
| $A_{2u}$ | 978 | (inactive) | |
| $B_{1u}$ | 1349 | 1375 | $\left(\begin{array}{c}\text{Absorption peaks of}\\\text{polyethylene}\end{array}\right)$ |
| $B_{2u}$ | 733<br>2928 | 728<br>2926 | |

## 8. The Sum Rule and the Product Rule for Rotational Isomers

In Section 6 we have explained in detail the procedure for calculating the normal frequencies of the 1,2-dihalogenoethanes and have shown how the vibrational frequencies of rotational isomers change with the azimuthal angle of internal rotation. From the secular equation (2.26) given in that section we can derive two fundamental rules governing the vibrational frequencies of rotational isomers. One of them was derived by Mizushima, Shimanouchi, Nakagawa and Miyake[8] and can be called the sum rule. The other which can be called the product rule was derived by Mizushima, Morino and Shimanouchi in 1942[9] and was also reported by Bernstein independently in 1949.[10]

From the secular equation in matrix notion:

$$|\boldsymbol{G}\,\boldsymbol{F} - \boldsymbol{E} \cdot 4\pi^2\,\nu^2| = 0 \qquad (2.26)$$

we have,

$$4\pi^2 \sum_{i=1}^{n} \nu_i{}^2 = \sum_{k,\,l=1}^{n} G_{k\,l}\,F_{k\,l} \qquad (2.38)$$

where $\nu_i'$ s are the normal frequencies and $n$ is the number of vibrational degrees of freedom ($n = 3\,N - 6$ or $3\,N - 5$, $N$ being the number of atoms in the given molecule). If we consider the interaction between the two movable groups to be small as compared with the other terms in the potential energy expression of the vibrational problem,* all the elements, $F_{k\,l}$, corresponding to $G_{k\,l}$ containing the azimuthal angle $\theta$ vanish, and the right-hand side of Eq. (2.38) becomes independent of the azimuthal angle $\theta$ of internal rotation. We have, therefore,

$$\sum_{i=1}^{n} \nu_i{}^2 = \text{constant.} \qquad (2.39)$$

---

* See, for example, Eq. (2.3). In this equation we have neglected the interaction between two chlorine atoms which are the two movable atoms in the skeleton of 1,2-dichloroethane.

---

8. S. Mizushima, T. Shimanouchi, I. Nakagawa and A. Miyake, J. Chem. Phys., **21**, 215 (1953).

9. S. Mizushima, Y. Morino and T. Shimanouchi, Sci. Pap. Inst. Phys. Chem. Res. Tokyo, **40**, 87 (1942).

10. H. J. Bernstein, J. Chem. Phys., **17**, 256 (1949).

This is the sum rule for rotational isomers. If, we apply this rule to the 1,2-dihalogenoethanes, we expect the sum of all the squared frequencies of the *trans* form to be equal to that of the *gauche* form. This has been shown to be actually the case for the dihalogenoethanes as shown in Table 2.7. Table 2.8 shows the results of the application of this rule to the skeletal frequencies of various dihalogenoethanes and halogenoacetyl halides. (For the former substances the skeletal frequencies used in these calculations are given in the same column as $\Sigma v_i^2$). In this case the difference in the sum of squared frequencies between the *trans* and the *gauche* molecules is larger than that in the case of the exact calculation, shown in Table 2.7, nevertheless it is still of significance to apply this rule for the confirmation of the assignment of skeletal frequencies.

TABLE 2.7

The sum of squared frequencies for the *trans* and the *gauche* isomers*

| Molecule | $\Sigma v_i^2$ (*Trans*) | $\Sigma v_i^2$ (*Gauche*) | Diff. (%) | Reference |
|---|---|---|---|---|
| $ClH_2C-CH_2Cl$ | $4.970 \times 10^7$ | $4.962 \times 10^7$ | 0.2 | (a), (b) |
| | $A_g$. 2957, 1445, 1304, 1052, 754, 300, $A_u$. 3005, 1124, 768. $B_g$. 3005, 1264, 989. $B_u$. 2957, 1450, 1230, 709, 223. | $A$. 2957, 1429, 1264, 1031, 654, 265, 3005, 1207, 943. $B$. 3005, 1145, 881, 2957, 1429, 1304, 677, 411. | | See also Table 2.4, Part I. |
| $BrH_2C-CH_2Br$ | $4.889 \times 10^7$ | $4.884 \times 10^7$ | 0.1 | (a), (b), (c) |
| $BrD_2C-CD_2Br$ | $2.807 \times 10^7$ | $2.810 \times 10^7$ | 0.1 | (c), (d) |

* The vibrational frequencies have been taken from the following papers:

(a) S. Mizushima, and Y. Morino, Proc. Ind. Acad. Sci., **8**, 315 (1938).

(b) T. Miyazawa, K. Kuratani and S. Mizushima, unpublished.

(c) J. T. Neu and W. D. Gwinn, J. Chem. Phys., **18**, 1642 (1950).

(d) S. Mizushima, Y. Morino and A. Suzuki, Sci. Pap. Inst. Phys. Chem. Res. Tokyo, **36**, 281 (1939).

TABLE 2.8

The sum of squared skeletal frequencies for the *trans* and the *gauche* isomers*

| Molecule | $\Sigma v_i^2$ (*Trans*) | $\Sigma v_i^2$ (*Gauche*) | Diff. (%) | Reference |
|---|---|---|---|---|
| $ClH_2C-CH_2Cl$ | $2.32 \times 10^6$ | $2.19 \times 10^6$ | 5.6 | (a) |
| | (1052, 754, 300, 709, 223.) | (1031, 654, 265, 677, 411.) | | |
| $BrH_2C-CH_2Br$ | $1.96 \times 10^6$ | $1.86 \times 10^6$ | 5.1 | (a) |
| | (1053, 660, 190, 589, 170.) | (1019, 551, 231, 583, 355.) | | |
| $ClH_2C-CH_2Br$ | $2.14 \times 10^6$ | $2.02 \times 10^6$ | 5.6 | (b) |
| | (1052, 726, 251, 630, 210.) | (1023, 665, 385, 568, 251.) | | |
| $ClH_2C-CH_2I$ | $2.01 \times 10^6$ | $2.11 \times 10^6$ | 5.0 | (b) |
| | (1046, 707, 232, 576, 186.) | (1107, 660, 368, 511. 232.) | | |
| $ClH_2C-COCl$ | $4.74 \times 10^6$ | $4.71 \times 10^6$ | 0.6 | (c) |
| $BrH_2C-COCl$ | $4.53 \times 10^6$ | $4.48 \times 10^6$ | 1.1 | (c) |
| $BrH_2C-COBr$ | $4.39 \times 10^6$ | $4.40 \times 10^6$ | 0.2 | (c) |

It is worthy of note that the sum rule can be applied satisfactorily to the *cis-trans* isomers of 1,2-dichloroethylene, in which the interaction between the two chlorine atoms is expected to be much larger than that in 1,2-dichloro-ethane. As shown in Table 2.9, the difference in $\Sigma v_i^2$ between the *trans* and *cis* forms is almost as small as that between the *trans* and *gauche* forms of 1,2-dichloroethane.

---

* The vibrational frequencies have been taken from the following papers:

(a) S. Mizushima and Y. Morino, Proc. Ind. Acad. Sci., **8**, 315 (1938).

(b) S. Mizushima and Y. Morino, Y. Miyahara, M. Tomura, and Y. Okamoto, Sci. Pap. Inst. Phys. Chem. Res. Tokyo, **39**, 387 (1942).

(c) I. Nakagawa, I. Ichishima, K. Kuratani, T. Miyazawa, T. Shimanouchi, and S. Mizushima, J. Chem. Phys., **20**, 1720 (1952).

TABLE 2.9

The sum of squared frequencies of the *trans* and *cis* isomers of 1,2-dichloroethylene*

| Molecule | $\Sigma v_i^2$ (*Trans*) | $\Sigma v_i^2$ (*Cis*) | Diff. (*) |
|----------|--------------------------|------------------------|-----------|
| ClHC = CHCl | $2.740 \times 10^7$ | $2.732 \times 10^7$ | 0.3 |
| ClDC = CDCl | $1.710 \times 10^7$ | $1.706 \times 10^7$ | 0.2 |

From the secular equation (2.26) we have

$$4\pi^2 \prod_{i=1}^{n} v_i^2 = |G|\,|F|.$$

If, therefore, we neglect the interaction between the two movable groups in the potential energy expression, the $F$-matrix becomes common to all rotational isomers and we have the following relation:

$$\frac{\prod\limits_{i=1}^{n} v_i'}{\prod\limits_{i=1}^{n} v_i} = \left(\frac{|G'|}{|G|}\right)^{1/2} \tag{2.40}$$

where $v_1$ and $v_1'$ are, respectively, the normal frequencies of two different rotational isomers. This is the product rule for rotational isomers.

The full lines of Fig. 2.5 represent the calculated values of the right-hand side of Eq. (2.40), or the square roots of ratios of $G'$ s of different azimuthal angles to those of the *trans* position at which the origin of the azimuthal angle $\theta$ has been taken. As can be seen from the figure the ratios of the products of the observed frequencies of 1,2-dichloroethane, 1,2-dibromoethane, ethylene chlorohydrin, chloroacetyl chloride and bromoacetyl chloride agree with the theoretical values at the azimuthal angles corresponding to the *gauche* forms.** Evidently, we do not mean by the *gauche* form a configuration with an azimuthal angle exactly equal to 120°. For example, the electron diffraction investigation of 1,2-dichloroethane (Section 8, Part I) showed this angle to be 109° ± 5°, while in the case of the halogenoacetyl halides this angle was found to be larger than the ideal value of 120°.

---

* The vibrational frequencies have been taken from the paper of H. J. Bernstein and D. A. Ramsay, J. Chem. Phys., **17**, 556 (1949).

** As shown in Fig. 2.5, the product rule can be applied fairly satisfactorily to the *cis-trans* isomers of ClHC=CHCl. As referred to above, the interaction between two Cl atoms in ClHC=CHCl is expected to be much larger than that in $ClH_2C–CH_2Cl$.

So far we have used the product rule for the determination of the azimuthal angle of the *gauche* form. As another example of the application of this rule, we shall determine the anti-symmetric deformation frequency of the skeleton of the *trans* form of 1,2-dichloroethane. This frequency is inactive in the Raman effect and is active in the infrared absorption, but it

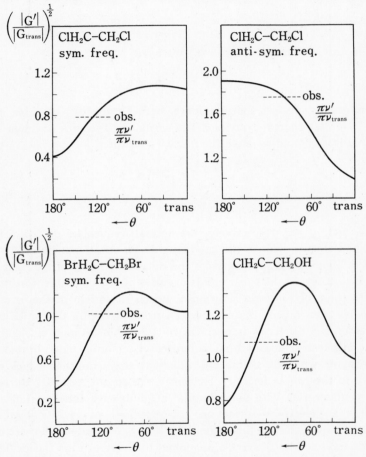

Fig. 2.5. The calculated values of $(|G'|/|G_{trans}|)^{1/2}$ and the observed values of $\Pi\ \nu'/\Pi\ \nu_{trans}$.

is not easy to observe this in the absorption spectrum, because this lies outside the observable region of ordinary infrared spectrometers. However as this is the only unknown frequency among all the skeletal frequencies of the *trans* and the *gauche* forms, we can calculate this frequency by the product rule.  Assuming the azimuthal angle of the *gauche* form to be 120°, we evaluate the right-hand side of Eq. (2.40) or the square root of the ratio of $G's$.  Then we put all other frequencies into the left-hand side and thus we can calculate the unknown frequency.  This has been shown to be close to 223 cm.$^{-1}$, at which a very weak Raman line has been observed in the

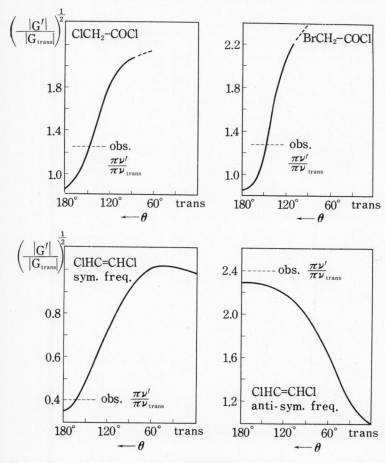

Fig. 2.5.  Continued.

liquid state.  This is the reason why we have assigned this Raman line to the anti-symmetric deformation vibration of the *trans* form which became active due to the deformation of the molecular configuration in the liquid state.  (See Section 4, Part I.)

It will not be out of place to discuss here on the dependence of skeletal frequencies on azimuthal angle. While some of these frequencies do not change much with azimuthal angle, others are very sensitive to the change of this angle in the region under consideration. For example, the calculated

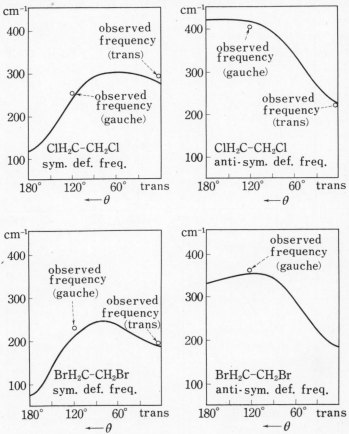

Fig. 2.6. Dependence of skeletal deformation frequencies on azimuthal angle.

value of the symmetric deformation frequency of the skeleton of $ClH_2C–CH_2Cl$ changes considerably with azimuthal angle in the region from $90°$ to $180°$ (Fig. 2.6). Therefore, we can determine fairly accurately the value of azimuthal angle which lies in the region referred to above. Similar relations

---

* The anti-symmetric deformation frequency of the *trans* form of $BrCH_2–CH_2Br$ has not been observed.

between the skeletal deformation frequency and the azimuthal angle is found for 1,2-dibromoethane, chloroacetyl chloride and bromoacetyl chloride. The calculated deformation frequencies are shown in Fig. 2.6 together with those observed (marked with small circles). We see from the figure that the results of these calculations are in good agreement with those obtained from the various experimental data given in Part I.

## 9. Assignment of the Raman and Infrared Frequencies of 1,2-Dichloroethane

A large number of experimental data have now been accumulated concerning the molecular structure of 1,2-dichloroethane, as described in Part I. They were obtained from the measurements of the Raman and infrared spectra in the gaseous, liquid and solid states and in solutions; dipole moments in the gaseous state and in solutions; electron diffraction in the gaseous state; X-ray diffraction in the solid state, and some thermal properties. All the data have been explained quite satisfactorily from the view that the molecules exist in the trans $(C_{2h})$ and gauche $(C_2)$ forms in the gaseous and liquid states and in solutions and they exist only in the trans form in the solid state. However, as to the vibrational frequencies, we have made the partial assignments, since so far we have calculated only the skeletal frequencies. Before concluding the discussion of rotational isomerism of the 1,2-dihalogenoethanes, we would like to make the complete assignment of the Raman and infrared frequencies of 1,2-dichloroethane based on the

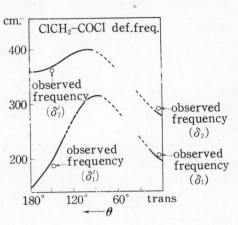

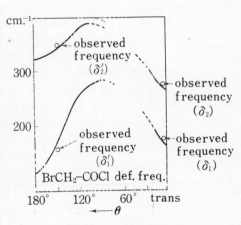

Fig. 2.6. Continued.

more exact calculation of normal vibrations recently made by Nakagawa of our laboratory.[11]

In this calculation the motions of all the atoms in the molecule are taken into account. In other words, the vibrations of $ClH_2C–CH_2Cl$ are treated as an eight-body problem. The potential function is of the Urey-Bradley

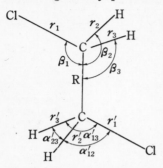

type with force constants shown in Table 2.10. The meaning of the notations for these force constants will be clear from the explanations given in Section 6. These force constants yield the calculated frequencies of the molecules with similar bonds (such as $CH_3Cl$, $CH_2Cl_2$, $CHCl_3$, $C_2H_6$, $C_2Cl_6$, etc.) which are in good agreement with those observed in the Raman effect and infrared absorption.[3] The values of these force constants are the same as those given in Table 2.5, but are different from those which we used in our calculation of skeletal vibrations in Section 6. As referred to in Section 7, this is due to the fact that in the latter case we have to include the interaction between the nonbonded carbon and hydrogen atoms in the force constants of the skeletons.

Fig. 2.7. Internal coordinates of $ClH_2C–CH_2Cl$ ($\alpha_{12}'$, $\alpha_{23}'$, $\alpha_{13}'$, $\beta_1'$, $\beta_2'$, and $\beta_3'$ which are not indicated in this figure have similar meaning as $\alpha_{12}'$, $\alpha_{23}'$, $\alpha_{13}'$, $\beta_1$, $\beta_2$ and $\beta_3$, respectively.)

We set up the kinetic energy matrix $G$ and the potential energy matrix $F$ by use of the internal coordinates shown in Fig. 2.7 and we replace the internal coordinates by the symmetry coordinates which correspond to displacements of the atoms in agreement with one of the symmetry classes.

TABLE 2.10

Force constants of 1,2-dichloroethane in $10^5$ dynes/cm.

| | | | |
|---|---|---|---|
| $K_{CC} = 2.80$, | $K_{CCl} = 1.81$, | $K_{CH} = 4.20$, | |
| $H_{ClCC} = 0.10$, | $H_{CCH} = 0.15$, | $H_{ClCH} = 0.05$, | $H_{HCH} = 0.40$, |
| $F_{CCl} = 0.60$, | $F_{CH} = 0.40$, | $F_{ClH} = 0.80$, | $F_{HH} = 0.10$. |
| $F' = -\dfrac{1}{10}F$, | | | |

($\varkappa = 0.10 \times 10^{-11}$ dynes $\cdot$ cm.)

11. I. Nakagawa, J. Chem. Soc. Japan, **74**, 848 (1953), **75** 178 (1954).

## TABLE 2.11
### Internal Symmetry Coordinates

| Class | Internal symmetry coordinates | Corresponding vibration |
|---|---|---|
| $A_g\ (C_{2h})$ <br> $A\ (C_2)$ | $\Delta s_1 = \Delta R$ | C–C stretching |
| | $\Delta s_2 = \dfrac{1}{\sqrt{2}}(\Delta r_1 + \Delta r_1')$ | C–Cl stretching |
| | $\Delta s_3 = \dfrac{1}{2}(\Delta r_2 + \Delta r_3 + \Delta r_2' + \Delta r_3')$ | C–H stretching |
| | $\Delta s_4 = \dfrac{1}{2\sqrt{6}}(\Delta \beta_2 + \Delta \beta_3 + \Delta \alpha_{12}$ <br> $+ \Delta \alpha_{13} - 2\Delta \beta_1 - 2\Delta \alpha_{23} + \Delta \beta_2'$ <br> $+ \Delta \beta_3' + \Delta \alpha_{12}' + \Delta \alpha_{13}' - 2\Delta \beta_1' - 2\Delta \alpha_{23}')$ | — |
| | $\Delta s_5 = \dfrac{1}{2\sqrt{2}}(\Delta \beta_2 + \Delta \beta_3 - \Delta \alpha_{12} - \Delta \alpha_{13}$ <br> $+ \Delta \beta_2' + \Delta \beta_3' - \Delta \alpha_{12}' - \Delta \alpha_{13}')$ | $CH_2$ wagging |
| | $\Delta s_6 = \dfrac{1}{2}(\Delta \beta_1 - \Delta \alpha_{23} + \Delta \beta_1' - \Delta \alpha_{23}')$ | — |
| $A_u\ (C_{2h})$ <br> $A\ (C_2)$ | $\Delta s_7 = \dfrac{1}{2}(\Delta r_2 - \Delta r_3 - \Delta r_2' + \Delta r_3')$ | C–H stretching |
| | $\Delta s_8 = \dfrac{1}{2\sqrt{2}}(\Delta \beta_2 - \Delta \beta_3 - \Delta \alpha_{12} + \Delta \alpha_{13}$ <br> $- \Delta \beta_2' + \Delta \beta_3' + \Delta \alpha_{12}' - \Delta \alpha_{13}')$ | $CH_2$ twisting |
| | $\Delta s_9 = \dfrac{1}{2\sqrt{2}}(\Delta \beta_2 - \Delta \beta_3 + \Delta \alpha_{12} - \Delta \alpha_{13}$ <br> $- \Delta \beta_2' + \Delta \beta_3' - \Delta \alpha_{12}' + \Delta \alpha_{13}')$ | $CH_2$ rocking |
| $B_g\ (C_{2h})$ <br> $B\ (C_2)$ | $\Delta s_{10} = \dfrac{1}{2}(\Delta r_2 - \Delta r_3 + \Delta r_2' - \Delta r_3')$ | C–H stretching |
| | $\Delta s_{11} = \dfrac{1}{2\sqrt{2}}(\Delta \beta_2 - \Delta \beta_3 - \Delta \alpha_{12} + \Delta \alpha_{13}$ <br> $+ \Delta \beta_2' - \Delta \beta_3' - \Delta \alpha_{12}' + \Delta \alpha_{13}')$ | $CH_2$ twisting |
| | $\Delta s_{12} = \dfrac{1}{2\sqrt{2}}(\Delta \beta_2 - \Delta \beta_3 + \Delta \alpha_{12} - \Delta \alpha_{13}$ <br> $+ \Delta \beta_2' - \Delta \beta_3' + \Delta \alpha_{12}' - \Delta \alpha_{13}')$ | $CH_2$ rocking |

| Class | Internal symmetry coordinates | Corresponding vibration |
|---|---|---|
| | $\Delta s_{13} = \dfrac{1}{\sqrt{2}} (\Delta r_1 - \Delta r_1')$ | C–Cl stretching |
| | $\Delta s_{14} = \dfrac{1}{2} (\Delta r_2 + \Delta r_3 - \Delta r_2' - \Delta r_3')$ | C–H stretching |
| $B_u (C_{2h})$ | $\Delta s_{15} = \dfrac{1}{2\sqrt{6}} (\Delta \beta_2 + \Delta \beta_3 + \Delta \alpha_{12} + \Delta \alpha_{13}$ $- 2 \Delta \beta_1 - 2 \Delta \alpha_{23} - \Delta \beta_2' - \Delta \beta_3' - \Delta \alpha_{12}'$ $- \Delta \alpha_{13}' + 2 \Delta \beta_1' + 2 \Delta \alpha_{23}'$ | — |
| $B (C_2)$ | $\Delta s_{16} = \dfrac{1}{2\sqrt{2}} (\Delta \beta_2 + \Delta \beta_3 - \Delta \alpha_{12} - \Delta \alpha_{13}$ $- \Delta \beta_2' - \Delta \beta_3' + \Delta \alpha_{12}' + \Delta \alpha_{13}')$ $\Delta s_{17} = \dfrac{1}{2} (\Delta \beta_1 - \Delta \alpha_{23} - \Delta \beta_1' + \Delta \alpha_{23}')$ | CH$_2$ wagging |

(see Table 2.11). Then we can factor the $G$- and $F$-matrices into those of lower degrees corresponding to each symmetry class. The order of each of these matrices which corresponds to the number of normal vibrations is shown in Table 2.12, together with the activity (selection rule) of each class of vibrations. As we do not include the torsional motion about the C–C axis in our calculation, the order of $A_u$ of the *trans* form becomes 3 instead of 4. and that of $A$ of the *gauche* form becomes 9 instead of 10 as shown in

TABLE 2.12

Number of normal vibrations belonging to each class of symmetry

| Class | Trans form ($C_{2h}$) | | | Class | Gauche form ($C_2$) | | |
|---|---|---|---|---|---|---|---|
| | number of vibrations | Activity | | | number of vibrations | Activity | |
| | | Raman | Infrared | | | Raman | Infrared |
| $A_g$ | 6 | + | — | $A$ | 10 | + | + |
| $A_u$ | 4 | — | + | | | | |
| $B_g$ | 3 | + | — | $B$ | 8 | + | + |
| $B_u$ | 5 | — | + | | | | |

Table 2.12. The values of normal frequencies $v_i$ can be calculated by solving the secular equation

$$|\boldsymbol{G}\,\boldsymbol{F} - \boldsymbol{E}\cdot 4\pi^2\,v^2| = 0$$

for each class. The calculated values are shown in Table 2.13, together with those observed in the Raman effect and infrared absorption.*

For the assignment of the observed frequencies we can make use of various experimental and theoretical methods described in the preceding chapters. First of all we apply the experimental technique of observing the spectral difference between the liquid and solid states, by which we can pick out separate spectra for the *trans* and the *gauche* forms. There remains, however, some uncertainties because of the overlapping by another line, etc. We then apply the sum rule derived in Section 8 in order to remove such uncertainties as far as possible. According to this rule, the sum of all the squared frequencies $(\Sigma v_i^2)$ should remain constant for the stable configurations of rotational isomers. From the observed frequencies shown in Table 2.13, we have

$$\sum v_i^2 \,(trans) = 4.970 \times 10^7$$

and

$$\sum v_i'^2 \,(gauche) = 4.962 \times 10^7.$$

Therefore, the assignment of observed frequencies to the *trans* and *gauche* forms shown in Table 2.13 is quite satisfactory.

After we have picked out separate spectra for the *trans* and the *gauche* forms, we have to assign each observed frequency to the corresponding normal vibration of one of these two stable forms. Evidently the first step is to consult the selection rule for the Raman and infrared spectra which is shown in Table 2.12 for each symmetry class. The state of polarization observed for a Raman line also helps us in making assignment, if it is an isolated intense Raman line. Unfortunately however, many lines of interest do not occur in this category. The complications of low intensity, the overlapping by other Raman lines or the exciting line, and weak polarization tend to make a decision difficult or impossible. Furthermore, a Raman line reported as depolarized is not necessarily ruled out of a "polarized class",

---

* The calculated values are in good agreement with those observed except for $v_{12}$ (1093 and 1264 cm.$^{-1}$). This seems to be due to the same origin as in the case of a frequency belonging to $B_{2g}$ of long chain paraffin molecules shown in Table 2.6 (1112 and 1295 cm.$^{-1}$).

for the depolarization degree of a totally symmetric vibration may take on any value up to the limiting value expected for a depolarized band.

Taking these into consideration as well as the calculated values of normal frequencies, we can assign the observed frequencies of Raman lines and the infrared absorption peaks as shown in Table 2.13. It is not necessary to describe the procedure of assignments for all the Raman lines and absorption peaks, but it is desirable to explain some of these assignments which do not seem to be straightforward.

TABLE 2.13

Calculated and observed frequencies in cm.$^{-1}$ of 1,2-dichloroethane

| Trans form: $C_{2h}$ | | | | Gauche form: $C_2$ | | | |
|---|---|---|---|---|---|---|---|
| Class | Mode | Calc. | Obs. | Class | | Calc. | Obs. |
| $A_g$ | $\nu_1$: C–H stretching | 2929 | 2957 | | $\nu_1'$ | 2931 | 2957 |
| | $\nu_2$: $CH_2$ bending | 1457 | 1445 | | $\nu_2'$ | 1454 | 1429 |
| | $\nu_3$: $CH_2$ wagging | 1294 | 1304 | | $\nu_3'$ | 1285 | 1264 |
| | $\nu_4$: C–C stretching | 1017 | 1052 | | $\nu_4'$ | 1019 | 1031 |
| | $\nu_5$: C–Cl stretching | 797 | 754 | | $\nu_5'$ | 671 | 654 |
| | $\nu_6$: CCCl bending | 279 | 300 | $A$ | $\nu_6'$ | 224 | 265 |
| $A_u$ | $\nu_7$: C–H stretching | 2967 | 3005 | | $\nu_7'$ | 2960 | 3005 |
| | $\nu_8$: $CH_2$ twisting | 1112 | 1124 | | $\nu_8'$ | 1098 | 1207 |
| | $\nu_9$: $CH_2$ rocking | 742 | 768 | | $\nu_9'$ | 917 | 943 |
| | $\nu_{10}$: torsional | — | — | | $\nu_{10}'$ | — | 125 |
| $B_g$ | $\nu_{11}$: CH stretching | 2957 | 3005 | | $\nu_{11}'$ | 2965 | 3005 |
| | $\nu_{12}$: $CH_2$ twisting | 1093 | 1264 | | $\nu_{12}'$ | 1108 | 1145 |
| | $\nu_{13}$: $CH_2$ rocking | 937 | 989 | | $\nu_{13}'$ | 862 | 881 |
| $P_u$ | $\nu_{14}$: C–H stretching | 2946 | 2957 | $B$ | $\nu_{11}'$ | 2943 | 2957 |
| | $\nu_{15}$: $CH_2$ bending | 1454 | 1450 | | $\nu_{15}'$ | 1456 | 1429 |
| | $\nu_{16}$: $CH_2$ wagging | 1300 | 1230 | | $\nu_{16}'$ | 1318 | 1304 |
| | $\nu_{17}$: C–Cl stretching | 737 | 709 | | $\nu_{17}'$ | 691 | 677 |
| | $\nu_{18}$: CCCl bending | 235 | 223 | | $\nu_{18}'$ | 396 | 411 |

A very weak Raman line at 223 cm.$^{-1}$ observed in our laboratory is assigned to the CCCl deformation frequency belonging to class $B_u$ of the *trans* form. This assignment apparently seems a violation of the selection rule, since class $B_u$ contains the normal vibrations antisymmetric to the center of symmetry which are forbidden in the Raman effect. However, this faint line is explained as arising from a molecule distorted in the liquid state by the influence of the neighboring molecules. The corresponding frequencies in $ClH_2C-CH_2Br$ or $ClH_2C-CH_2I$[12] become Raman active, since we have no center of symmetry for the *trans* form of these molecules. It can be shown that the extrapolation from these observed frequencies yields a value of the $ClH_2C-CH_2Cl$ frequency close to 223 cm.$^{-1}$. The product rule has also supported this assignment, as explained in Section 8. Evidently it is desirable to measure this frequency directly as an infrared absorption frequency, but the frequency value is too low to be measured by a usual prism spectrometer*.

The Raman line at 125 cm.$^{-1}$ should be assigned to a vibration of the *gauche* form, since this has been observed only in the liquid state. For this frequency we have made no calculation such as we did for other frequencies,** but we have good reason to believe in assigning this frequency to the torsional motion about the C–C axis. First, this frequency lies far below all other frequencies calculated for this molecule and second this accounts for the entropy value contributed by the torsional motion. As stated in Section 9, Part I this is obtained by subtracting from the thermal entropy, the molecular entropy in which all other degrees of freedom are taken into account. The torsional frequency of the *trans* form is inactive in the Raman effect and is too low to be observed by an usual infrared apparatus.

A note may be added on the coupling of the torsional motion with the CCCl deformation frequency. In the *trans* form there is no coupling between them, since the former $(\nu_{10})$ belongs to class $A_u$, while the latter $(\nu_6)$ to class $A_g$. In the *gauche* molecule, however, both of them $(\nu_{10}'$ and $\nu_6')$ belong to the same class, $A$, and accordingly, there is a coupling between them. This means that the pure torsional motion has a frequency higher than that observed (125 cm.$^{-1}$).***

---

*After completing this manuscript, the author noticed a paper by C. R. Bonn, N. K. Freeman, W. D. Gwinn, J. L. Hollenberg and K. S. Pitzer, J. Chem. Phys. **21**, 719 (1953) in which these researchers reported an absorption band at 222 cm$^{-1}$ observed by a grating spectrometer. This value is in excellent agreement with the value 223 cm$^{-1}$ referred to above.

** In the potential function of $ClH_2C-CH_2Cl$ we have not included azimuthal angle as an internal coordinate and have neglected the interaction between the two movable groups and, therefore, the torsional frequency can not be calculated.

*** According to Morino and Shimanouchi, the pure torsional frequency may be higher than that observed by 30 cm.$^{-1}$.

---

12. S. Mizushima, Y. Morino, Y. Miyahara, M. Tomura and Y. Okamoto, Sci. Pap. Inst. Phys. Chem. Res. Tokyo, **39**, 387 (1942).

INTERNAL ROTATION. PART II

A similar coupling problem may arise between the C–Cl stretching $(\nu_5, \nu_5')$ and the $CH_2$ rocking frequencies $(\nu_9, \nu_9')$. In the *trans* form for which $\nu_5$ belongs to class $A_g$ and $\nu_9$ belongs to class $A_u$, there is no coupling, while in the *gauche* form for which both $\nu_5'$ and $\nu_9'$ belong to the same class, $A$, there is a coupling between them. The fact that $\nu_5'$ of the *gauche* form (654 cm.$^{-1}$) is lower than $\nu_5$ of the *trans* form (754 cm.$^{-1}$) while $\nu_9'$ (943 cm.$^{-1}$) is much higher than $\nu_9$ (768 cm.$^{-1}$) is considered to be partly due to this situation.*

In order to discuss such a problem rigorously, we have to calculate the relative amplitude for each of the internal coordinates (e. g. change of bond lengths or bond angles) or internal symmetry coordinates in a given normal mode of vibration.** This is done in the following way.

The internal coordinates, $R$, or the internal symmetry coordinates, $S$, are related linearly to the normal coordinates, $Q$, through the matrix expressions:

$$R = L\,Q$$

and

$$S = L'\,Q$$

If, therefore, we calculate the elements of the $L$-matrix or of the $L'$-matrix for a given normal vibration, we can at once determine the relative amplitude for each internal coordinate or for each symmetry coordinate. Such calculations have been made for all the normal vibrations of 1,2-dichloroethane and the result is shown in the second column of Table 2.13 as C–H stretching, $CH_2$ bending, etc. We must, however, realize that these are very rough representations, since none of the vibrations is restricted in a single internal coordinate or in a single internal symmetry coordinate as these terms might suggest.

We have stated above that the observed frequency 754 cm.$^{-1}$ can be assigned to the symmetric C–Cl stretching vibration of the *trans* form which is not coupled with $CH_2$ rocking frequency. By calculation of the $L$-matrix we can also show that other kinds of hydrogen vibrations play no significant part in this C–Cl vibration.[11] However, we see from the same calculation that there is a large contribution of the CCCl bending vibration and

---

* We cannot attribute the whole difference between $\nu_5$ and $\nu_5'$ or between $\nu_9$ and $\nu_9'$ to the coupling effect. A part of this difference is due to the change of molecular configuration with the change of azimuthal angle.

** It will easily be seen that in many cases we have a better picture of the vibration in internal symmetry coordinate than in usual internal coordinate.

accordingly, we must conclude that in this vibration there is a conspicuous change not only in C–Cl bond length but also in CCCl bond angle, although we designate it as C–Cl stretching vibration. In any event as the coupling in this vibration takes place only among the motions of heavier atoms, the calculation of skeletal vibration is a good approximation for the solution of the problem of normal vibrations.

By contrast the calculation of the $L$-matrix for the corresponding vibration of the *gauche* form $(\nu_5')$ shows that the contribution of motions of hydrogen atoms is considerable in this case, as has been expected. This is due to the fact that some hydrogen deformation vibrations belong to the same symmetry class $(A)$ as $\nu_5'$. Among these hydrogen deformation vibrations, $\nu_9'$ ($CH_2$ rocking vibration) is most important in the coupling with $\nu_5'$ as stated above. $CH_2$ wagging vibration also makes a considerable contribution to this normal vibration. Therefore in such a normal vibration, the calculation of skeletal vibration in which the $CH_2$-group is treated as a mass point, cannot be considered to be a good approximation.*

In connection with these vibrational problems, a few lines may be added as to the assignment of the C–C stretching frequency $(\nu_4)$ of the *trans* form. This was first assigned by Mizushima and Morino[13] to the Raman line at 1052 cm.$^{-1}$, but afterwards, some investigators preferred the 989 cm.$^{-1}$ line. The same problem has arisen as to the assignment of the C–C stretching frequency of 1,2-dibromoethane which was first assigned by Mizushima and Morino to the 1053 cm.$^{-1}$ line but some investigators preferred the 933 cm.$^{-1}$ line to this line, because the 1053 cm.$^{-1}$ line was reported depolarized, which would not be the case of a line belonging to class $A_g$ containing only totally symmetric vibrations. However, as to the polarization of a Raman line there is a problem mentioned above, and, therefore, the reported depolarization of the 1053 cm.$^{-1}$ line is not a direct contradiction of the assignment of this line to the C–C stretching frequency. Neu and Gwinn[14] who made a thorough investigation of the vibrational spectrum of 1,2-dibromoethane also assigned the 1053 cm.$^{-1}$ line to the C–C stretching frequency. As to the

---

* We should like to note that the contribution of hydrogen vibration is small in the case of CCCl deformation vibrations of both the *trans* and *gauche* forms. Therefore, the calculations of skeletal deformation vibrations described in the last part of Section 8 can be considered to be a good approximation of the solution of the problem.

---

13. S. Mizushima and Y. Morino, Sci. Pap. Inst. Phys. Chem. Res. Tokyo, **29**, 188 (1936).

14. J. T. Neu and W. D. Gwinn, J. Chem. Phys., **18**, 1642 (1950).

polarization of the 1052 cm.$^{-1}$ line of 1,2-dichloroethane there is no question, since Neu, Ottenberg and Gwinn[15] reported this line to be polarized. However, the calculation of the $L$-matrix for the computed frequency $\nu_4$, which is close enough to the observed line at 1052 cm.$^{-1}$ in frequency, settles this problem definitely. This calculation shows that we can properly designate $\nu_4$ as the C–C stretching frequency in which the vibration takes place mainly in the C–C bond. The alternative line at 989 cm.$^{-1}$ is close to the calculated frequency, $\nu_{13}$, of the $CH_2$ rocking motion.

In the discussions given above we have dealt with the frequencies which are lower than the C–C stretching frequencies. The frequencies higher than these are all assigned to the hydrogen vibrations, either stretching or deformation. Of these hydrogen vibrations, the C–H stretching vibrations are expected to be simple in character, because these frequencies appear in a region (about 3000 cm.$^{-1}$) far apart from those of other frequencies, so that the coupling problem does not come into question. Actually the calculation of the $L$-matrix made for various C–H stretching frequencies, $\nu_1$ $(A_g)$, $\nu_7$ $(A_u)$, $\nu_{11}$ $(B_g)$ and $\nu_{14}$ $(B_u)$, shows that the vibration of large amplitude takes place only in the C–H bond. The other group of hydrogen vibrations appears in the frequency region below 1500 cm.$^{-1}$ and contains $CH_2$ bending, $CH_2$ rocking, $CH_2$ wagging and $CH_2$ twisting vibrations. Their vibrational modes are shown in Fig. 2.8. In all of them the hydrogen atom moves in a direction perpendicular to C–H bond and this is the reason why these frequencies are much lower than hydrogen stretching frequencies. The calculation of the $L$-matrix for these frequencies shows that they are considerably pure frequencies: in other words the actual modes of these normal vibrations are expressed fairly accurately by the displacement vectors shown in Fig. 2.8. One might consider fairly strong coupling of the C–C stretching frequency with some of these hydrogen frequencies, but as stated above, the calculation of the $L$-matrix for these hydrogen frequencies shows that such is not actually the case. We have already shown that $\nu_4$ is also fairly pure C–C stretching frequency.

The assignments explained above all refer to the fundamental frequencies. Besides these frequencies, there have been observed (especially in the infrared absorption spectra) some overtones and combination tones. In the spectra observed in the solid state, there appear some low frequency Raman lines to which we have referred in Section 8, Part I. These lines can be assigned neither to the fundamental frequencies nor to the overtones of intra-

15. J. T. Neu, A. Ottenberg and W. D. Gwinn, J. Chem. Phys., **16**, 1004 (1948).

molecular vibrations, since their frequencies are too low to be accounted for as such (53 and 74 cm.$^{-1}$ at $-140°$ C and 66 cm.$^{-1}$ at $-40°$ C). These lines should be assigned to the lattice vibrations of the crystal which can be considered to be Raman active from the selection rule for the crystals. This rule has been developed by Halford[16] and also by Ichishima[17] and is applied to the present case as follows.

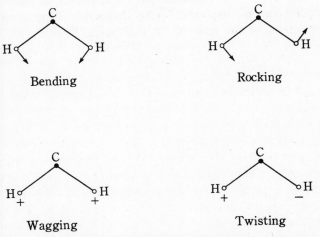

Bending     Rocking

Wagging     Twisting

Fig. 2.8. The modes of $CH_2$ bending, rocking, wagging and twisting vibrations. ($+$ and $-$ signs indicate motions perpendicular to the plane of the paper.)

The molecules of 1,2-dichloroethane take the *trans* form in the crystalline state. We can consider approximately that a molecule in the crystal is moving in a potential field formed by surrounding molecules in their equilibrium positions with a symmetry corresponding to one of the subgroups of $C_{2h}$ which is the symmetry of the *trans* molecule. The existence of the center of symmetry (one of the subgroups of $C_{2h}$) which is very probable in this case allows only the rotatory vibrations in the Raman effect and forbids the translatory vibrations. Of the three Raman active rotatory vibrations, one about the molecular axis would escape detection owing to the small change of polarizability. Therefore, the two frequencies 53 and 74 cm.$^{-1}$ should be assigned to the rotatory vibrations about the two axes,

16. R. S. Halford, J. Chem. Phys., **14**, 8 (1946).
17. I. Ichishima, Rep. Rad. Chem. Res. Inst. (Tokyo University), **4**, 9 (1949).
See also I. Ichishima and S. Mizushima, J. Chem. Phys., **18**, 1420 (1950).

both perpendicular to the molecular axis.　Instead of these two lines we have observed only one at higher temperatures.　As explained in Section 8, Part I, this is due to the fact that above the transition point the free rotation about the molecular axis* sets in and there is no more difference between the two rotatory vibrations stated above.

The two Raman lines of crystalline 1,2-dibromoethane observed at 41 and 53 cm.$^{-1}$ below the transition point and one at 49 cm.$^{-1}$ above the transition point can be explained quite similarly.　They arise from the rotatory vibrations of the molecule in the crystal lattice.

---

* Evidently this is the rotation of the molecule as a whole, and should not be confused with the internal rotation about the C–C axis.

# Author Index

# Subject Index

232

## K

α-Keratin, 140
and synthetic polypeptide, 144
backbone spacing, 147
conversion to β, 144, 147
folded chain, 141
molecular model, 140, 141
side-chain spacing, 147
structure, 147, 148
wool, 149
β-Keratin, backbone spacing, 147–149
fibre diagram, 139, 140
from α, 146–149
oriented film, 150

## L

Lattice vibration, 225
Lead sulfide photocell, 159
Lead telluride, 159
Lennard-Jones interaction, 56
Light scattering, 162
Liquid, Raman tubes, 167, 168
Liquid, dielectric, Debye's theory, 83
Liquid absorption cell, 160
Liquid state, and molecular distortion, 20
molecular interaction, 43, 44
rotational isomerism, 25
L-Matrix, 222
Low temperature absorption cell, 161

## M

Melting point, convergence of, 111
homologous series, 110
Mercury arc, 164, 165
Mercury dimethyl, 55
Meso form, 15
Meso tartaric acid, 14
N-Methylacetamide, 118
absorption peak frequencies, 123
absorption spectra, 121
association, 120
and bonding, 123
carbon-oxygen stretching frequency, 123
intermolecular hydrogen bond, 122, 125
molecular absorption coefficient, 128
molecular association, 120, 122, 129, 130

N-Methylacetamide,
near infrared absorption, 122, 123
resonance, 134
structure and hydrogen bonding, 128
skeletal frequency, 119
N-Methylacetamidionium ion, 118, 119
Methyl acetate, dipole moment, 89
Methyl alcohol, carbon-oxygen bond, 83
heat capacity, 85
potential barrier to internal rotation, 84
2-Methylbutane, 37
Methyl chloroform, 57
Methyl chloroformate, 89
dipole moment, 89, 90
molecular configuration, 90
Methyl cyclohexane, 78, 79
Methyl halide, 10, 11
Methyl oxalate, 74
stable configuration, 75
2-Methylpentane, 100
3-Methylpentane, 100
Microwave spectroscopy, 84
Molecular absorption coefficient, 39, 125, 161
Molecular entropy, 51, 52, 67
n-butane, 98
1,2-dichloroethane, 52
methyl alcohol, 84
Molecular polarization, 170
and dipole moment, 172
Molecular refraction, and refractive index, 171
Molecular scattering term, 180
Molecule, carbonyl containing, 73, 74
carboxylic acid configuration, 88
configuration, and covering operations, 31
specificity, 152
cyclohexane, 77, 80
gauche form, 22
trans form, 20
distortion, 20
eclipsed form, 29
frequency, empirical method of assignment, 187
improper rotation, 31
interaction, in liquid state, 43, 44